UNIVERSAL FORESTRY

2nd EDITION

UNIVERSAL FORESTRY

2nd EDITION

(UPSC, PCS, ARS/SRF/JRF/AFO, STATE PG & PH.D. ENTRANCE EXAMINATIONS AND INTERVIEWS OF ALL FOREST SERVICES)

Mohit Husain

M. Sc. Forestry (Agroforestry and Watershed Management)

SKUAST- Kashmir

(UG-Gold Medalist)

Published by
SCIENTIFIC PUBLISHERS (INDIA)

Jodhpur –

5 A, New Pali Road
P.O. Box 91
Jodhpur - 342 001 INDIA

Print 2026

ISBN: 978-93-89184-24-2
eISBN: 978-93-89184-25-9

Printed in India

Dedicated to my beloved parents

Mr. Amisher Khan

&

Mrs. Hanipha Khatoon

Preface

Forestry is a broad subject. It covers all the basic subject of science. Therefore students have to pass through all the forestry related subject with elementary courses during under graduation. After passing graduation, higher degrees are done in specialized field of forestry, so there is no need to read all forestry subject in M.Sc. and Ph.D. programme.

But for the preparation of various forestry competition examinations students have to read all the basic books of forestry to cover the syllabus. And at that time most of the students do not have all the necessary books and ample of time to read them.

Therefore, to overcome these problems the author have tried his best to write this book through reading various books, states forest reports and other sources of forestry, to covers brief information of all the subjects in one book.

Covering all the subjects of forestry in one books is a very hard task so some errors may be present, therefore your suggestions are invited (mohit.husainchf@gmail.com) for further improvement of this book.

I express my appreciation for the hard work and affinity of my sincere and respected professors who have guided me for bring all the content in this single book.

The forbearance shown by my mother, Mrs. Hanipha Khatoon, father Mr. Amisher Khan, brothers Mr. Kamil Hussain & Mr. Suhaiv Akhtar and sister Ms. Nargis Khatoon, is worth placing high on record who always favoured me in collecting and bearing the assignment.

I hope this book would be very helpful for those appearing in various competitive exams of forestry.

Mohit Husain

About the Author

Mr. Mohit Husain was born in Dhoondli (Bijnor), a small village in the state of Uttar Pradesh. His father, Amisher Khan is a businessman and motivated him to pursue a under graduation in forestry science. Mr. Mohit Husain completed B.Sc. (Forestry) from college of Horticulture and Forestry, Agricultural University Kota, Rajasthan in 2016 and Master's degree in forestry science from Sher-e-Kashmir University of Agricultural Sciences and Technology of Kashmir in 2018. He has qualified ICAR-UG, ICAR-PG and ASRB-NET (Agroforestry) examinations.

He was awarded *gold medal* and *Best Student of the Year-2016* during his graduation. He was also awarded as *Young Scientist Associated award* by Agricultural Technology Development Society, Ghaziabad, Uttar Pradesh. He has received *Best Master Thesis award* and *Best Poster presentation* award in international conference.

Mr. Mohit Husain attended various National and International conferences/ seminars and gave oral presentations as well as poster presentations.

He has done various trainings in both Agricultural and Forestry field. He has published more than thirty research, review and popular articles in National and International journals.

He had published another book entitled as *Universal Objective Forestry* (ISBN: 9789387307926) in which previous years all forestry as well as forestry allied exam papers are included.

He has also written a practical manual on Tree physiology, Forest nursery and also edited books, entitled *Objective Forestry* and *Sal Mortality and Mitigation*.

JRF Syllabus for Forestry

UNIT I: Importance of Agriculture/Forestry/Livestock in national economy. Basic principles of crop production. Important rural development programmes in India Elementary principles of economics and agri-extension. Organizational set up of Agricultural Research, education and extension in India. Major diseases and pests of crops. Elements of statistics.

UNIT II: Forest-importance, types, classification, ecosystem, biotic and abiotic components, ecological succession and climax, nursery and planting technique, social forestry, farm forestry, urban forestry, community forestry, forest management, silvicultural practices, forest mensuration, natural regeneration, man-made plantations, shifting cultivation, taungya, dendrology, hardwoods, softwoods, pulp woods, fuel woods, multipurpose tree species, wasteland management. Agro forestry – importance and land use systems, forest soils, classification and conservation, watershed management, forest genetics and biotechnology and tree improvement, tree seed technology, rangelands, wildlife – importance, abuse, depletion, management, major and minor forest products including medicinal and aromatic plants, forest inventory, aerial photo interpretation and remote sensing, forest depletion and degradation – importance and impact on environment, global warming, role of forests and trees in climate mitigation, tree diseases, wood decay and discolouration, tree pests, integrated pest and disease management, biological and chemical wood preservation, forest conservation, Indian forest policies, Indian forest act, forest engineering, forest economics, joint forest management and tribology.

NET Syllabus for Agroforestry

UNIT 1. National Forest Policy 1894, 1952 and 1988; Indian Forest Act, 1927; Forest Conservation Act, 1980 and Wildlife Protection Act, 1972; Forests-extent, basis for classification and distribution in India; Geographical distribution and salient features of major world forest types; Phytogeographical regions and vegetation of India; Role of forests in national economy - productive, protective and ameliorative, tribal and rural livelihoods; Forest types of India: distribution and types; Succession, climax and retrogression; Concepts of biomass, productivity, energy flow and nutrient cycling in forest ecosystem; Migration and dispersal mechanism.

UNIT 2. Concept and definition of agroforestry, social forestry, community forestry and farm forestry; Benefits and constraints of agroforestry; Historical development of agroforestry and overview of global agroforestry systems. Classification of agroforestry systems: structural, functional, socio-economic and ecological; Diagnosis and design of agroforestry system; Land capability classification and land use; Criteria of an ideal agroforestry design, productivity, sustainability and adoptability; multipurpose tree species and their characteristics suitable for agroforestry.

UNIT 3. Plant management practices in agroforestry; Tree-crop interactions: ecological and economic; Concept of complementarity, supplementarity and competition; Productivity, nutrient cycling and light, water and nutrient competition in agroforestry; Concept of allelopathy and its impact on agroforestry; Energy plantations - choice of species and management; Lopping of top-feed species such as frequency and intensity of lopping; Organic farming; Financial analysis and economic evaluation of agroforestry systems: cost benefit analysis and land equivalent ratio; Agroforestry practices and systems in different agro - ecological zones of India.

UNIT 4. Extent and causes of land denudation; Effects of deforestation on soil erosion, land degradation, environment and rural economy; Wastelands: their extent, characteristics and reclamation; Watershed management and its role in social, economic and ecological development; Biomass production for fuel wood, small timber, raw material for plant-based cottage industries, non-wood forest products such as gums, resins & tannins, medicinal plants, essential oils, edible fruits, spices, bamboo and canes; Wood quality and wood preservation; Plywood and pulp industries.

UNIT 5. Forest mensuration - definition, object and scope; Measurement of diameter, girth, height, stem form, bark thickness, crown width and crown length; Measurement methods and their principles. Measurement and computation of volume of logs and felled/standing trees; Construction and application of volume tables; Biomass measurement; Growth and increment; Measurement of crops; Forest inventory: kinds of enumeration, sampling methods, sample plots and photo interpretation; Geographic information systems and remote sensing - concept and scope.

UNIT 6. Definition, object and scope of silviculture; Site factors - climatic, edaphic, physiographic, biotic and their influence on forest vegetation; Forest regeneration: natural and artificial; Silvicultural systems - high forest and coppice systems; Silviculture of important tree species - Populus, Eucalyptus, Dalbergia, Acacia, Tectona, Shorea, Prosopis, Casurina, Pinus, Gmelina, Azadirachta, Diospyros, Pterocarpus, Anogeissus, Santalum, Quercus and Albizia.

UNIT 7. Seed collection, processing, storage, viability and pre-treatment; Seed dormancy and methods for breaking dormancy; Seed testing and germination tests; Seed certification and ISTA Rules; Forest nursery - need, selection and preparation of site, layout and design of nursery beds; Types of containers; Root trainers; Growing media and sowing methods; Management of nursery-shading, watering, manuring, fertilizer application, weed control, insect pest and diseases control; Planting techniques: site selection, evaluation and protection; Soil working techniques for various edaphic and climatic conditions; Planting patterns; Plant spacing, manure and fertilizer application, irrigation/moisture conservation techniques; Choice of species. Afforestation on difficult sites: saline-alkaline soils, coastal sands, lateritic soils, wetlands, ravines and sand dunes, dry and rocky areas, cold desert; Tending operations - weeding, cleaning, climber cutting, thinning - mechanical, ordinary, crown and selection thinning, improvement felling, pruning and girdling; Forest fires: causes, types, impacts and control measures; Major forest pests and weeds.

UNIT 8. Forest management: definition and scope; Concept of sustained yield and normal forest; Rotation; Estimation of growing stock, density and site quality; Management of even aged and uneven aged forest; Regulation of yield in regular and irregular forests by area, volume, increment and number of trees; Working plan; Joint forest management; Conservation and management of natural resources including wildlife; Forest evaluation; Internal rate of return, present net worth and cost benefit analysis.

UNIT 9. Tree improvement: nature and extent of variations in natural population; Natural selection; Concept of seed source/ provenance; Selection of superior trees; Seed production areas, exotic trees, land races; Collection, evaluation and maintenance of germplasm; Provenance testing. Genetic gains; Tree breeding: general principles, mode of pollination and floral structure; Basics of forest genetics - inheritance, Hardyweinburg Law, genetic drift; Aims and methods of tree breeding. Seed orchard: types, establishment, planning and management, progeny test and designs; Clonal forestry - merits and demerits; Techniques of vegetative propagation, tissue culture, mist chamber; Role of growth substances in vegetative propagation.

UNIT 10. Forestry in bio-economic productivity of different agro-eco-systems and environmental management; Global overview and classification of agroforestry systems; Tree-crop interaction in agroforestry; Biomass production for fuel' wood, small timber, raw material for plants-based cottage industries, non-wood forest products such as gums, resins, tannins, medicinal plants, essential oils, edible fruits, bamboos and canes; Principle and criteria of plant selection in agroforestry; Resource use-efficiency in agroforestry.

UNIT 11. Measurement of trees and stand – diameter, girth, height, form and crown characteristics; Measurement methods and their principles; Volume/biomass estimation, volume tables; Measurement of rangeland productivity; Forest enumeration: sampling methods, sample plots, surveys and photo interpretation; Concept and application of GIS and remote sensing; Introduction to internal rate of return, present net worth, cost benefit analysis and land equivalent ratio; Agroforestry and environmental conservation; Role of green revolution in forest conservation in India.

UNIT 12. Climate change: greenhouse effect, sources and sinks of green-house gases, major greenhouses gases; Global climate change – its history and future predictions; Impact of climate change on agriculture, forestry, water resources, sea level; Livestock, fishery and coastal ecosystems; International conventions on climate change; Global warming: effect of enhanced CO_2 on productivity; Ozone layer depletion; Disaster management, floods, droughts, earthquakes; Tsunami, cyclones and landslides; Agroforestry and carbon sequestration.

UNIT 13. Statistics: definition, object and scope; Frequency distribution; Mean, median, mode and standard deviation, introduction to correlation and regression; Experimental designs: basic principles, completely randomized, randomized block, Latin square and split plot designs.

Contents

CHAPTER 1

General Forestry

STATE WISE FOREST PROFILE

1. ANDHRA PRADESH	
Total Geographic Area	160,204 Km^2
Recorded forest area	3,965 Km^2
Forest area of TGA	2.48 per cent
Annual Rainfall	1094 mm
Total Protected areas	16
National Parks	3
Wildlife Sanctuaries	13
State animal	Blackbuck
State bird	Indian roller
State tree	Neem
Local tribes	Andh, Bhil, Gond, Konda
2. ARUNACHAL PRADESH	
Total Geographic Area	83,743 Km^2
Recorded forest area	761 Km^2
Forest area of TGA	0.91 per cent
Annual Rainfall	2782 mm
Total Protected areas	13
National Park	2
Wildlife Sanctuary	11
State animal	Mithun
State bird	Great Hornbill
State tree	Hollong
Local tribes	Abor, Galong, Khowa, Bhil

3. ASSAM	
Total Geographic Area	78,438 Km^2
Recorded forest area	1,613 Km^2
Forest area of TGA	2.06 per cent
Annual Rainfall	2818 mm
Total Protected areas	23
National Park	5
Wildlife Sanctuary	18
State animal	One horn Rhinoceros
State bird	White-winged wood Duck
State tree	Hollong
Local tribes	Chakma, Garo, Hajong, Kuki
4. BIHAR	
Total Geographic Area	94,163 Km^2
Recorded forest area	2,182 Km^2
Forest area of TGA	2.32 per cent
Annual Rainfall	1250 mm
Total Protected areas	13
National Park	1
Wildlife Sanctuary	12
State animal	Ox
State bird	House sparrow
State tree	Peepal
Local tribes	Asar, Bhumj, Ho, Gond, Kisan
5. CHHATTISGARH	
Total Geographic Area	1,35,191 Km^2
Recorded forest area	3,629 Km^2
Forest area of TGA	2.68 per cent
Annual Rainfall	1292 mm
Total Protected areas	14
National Park	3
Wildlife Sanctuary	11
State animal	Wild Buffalo
State bird	Hill myna

State tree	Sal
Local tribes	Agariya, Andh, Baiga, Pajra
6. GOA	
Total Geographic Area	3,702 Km^2
Recorded forest area	325 Km^2
Forest area of TGA	8.78 per cent
Annual Rainfall	3005 mm
Total Protected areas	7
National Park	1
Wildlife Sanctuary	6
State animal	Gaur
State bird	Ruby Throated yellow Bulbul
State tree	Matti
Local tribes	Dhodia, Dubla, Siddi, Varli
7. GUJARAT	
Total Geographic Area	1,96,022 Km^2
Recorded forest area	7,914 Km^2
Forest area of TGA	4.04 per cent
Annual Rainfall	1107 mm
Total Protected areas	27
National Park	4
Wildlife Sanctuary	23
State animal	Asiatic Lion
State bird	Greater Flammingo
State tree	Mango
Local tribes	Baeda, Dhodia, Gond, Koli, Patelia
8. HARYANA	
Total Geographic Area	44,212 Km^2
Recorded forest area	1,355 Km^2
Forest area of TGA	3.07 per cent
Annual Rainfall	617 mm
Total Protected areas	10
National Park	2
Wildlife Sanctuary	8

State animal	Blackbuck
State bird	Black Francolin
State tree	Peepal
Local tribes	
9. HIMANCHAL PRADESH	
Total Geographic Area	55,673 Km^2
Recorded forest area	757 Km^2
Forest area of TGA	1.36 per cent
Annual Rainfall	1251 mm
Total Protected areas	33
National Park	5
Wildlife Sanctuary	28
State animal	Snow Leopard
State bird	Western Tragopan
State tree	Himalayan Ceder
Local tribes	Gaddi, Laholi, Pangwal, Bholia
10. JAMMU & KASHMIR	
Total Geographic Area	2,22,236 Km^2
Recorded forest area	8,354 Km^2
Forest area of TGA	3.76 per cent
Annual Rainfall	1011 mm
Total Protected areas	19
National Park	4
Wildlife Sanctuary	15
State animal	Hangul
State bird	Black necked Crane
State tree	Chinar
Local tribes	Gujjar, Bakarwal
11. JHARKHAND	
Total Geographic Area	79,714 Km^2
Recorded forest area	2,783 Km^2
Forest area of TGA	3.49 per cent
Annual Rainfall	1543 mm
Total Protected areas	12

National Park	1
Wildlife Sanctuary	11
State animal	Elephant
State bird	Koel
State tree	Sal
Local tribes	Asur, Munda, Oraon, Santhal
12. KARNATAKA	
Total Geographic Area	1,91,791 Km^2
Recorded forest area	5,552 Km^2
Forest area of TGA	2.89 per cent
Annual Rainfall	3456 mm
Total Protected areas	32
National Park	5
Wildlife Sanctuary	27
State animal	Indian Elephant
State bird	Indian roller
State tree	Sandal
Local tribes	Adiyan, Toda, Varli
13. KERALA	
Total Geographic Area	38,863 Km^2
Recorded forest area	2,951 Km^2
Forest area of TGA	7.59 per cent
Annual Rainfall	3055 mm
Total Protected areas	23
National Park	6
Wildlife Sanctuary	17
State animal	Elephant
State bird	Great Hornbill
State tree	Coconut
Local tribes	Chenchu, Irula, Sumali, Toda
14. MADHYA PRADESH	
Total Geographic Area	3,08,245 Km^2
Recorded forest area	7,773 Km^2
Forest area of TGA	2.52 per cent

Annual Rainfall	1338 mm
Total Protected areas	34
National Park	9
Wildlife Sanctuary	25
State animal	Swamp Deer
State bird	Asian Paradise Flycatcher
State tree	Banyan
Local tribes	Bhil, Todiya, Kol, Bhumia
15. MAHARASHTRA	
Total Geographic Area	3,07,713 Km^2
Recorded forest area	9558 Km^2
Forest area of TGA	3.11 per cent
Annual Rainfall	1034 mm
Total Protected areas	47
National Park	6
Wildlife Sanctuary	41
State animal	Indian Giant Squirrel
State bird	Green Pigeon
State tree	Mango
Local tribes	Andh, Barda, Bihor, Bhil, kamar
16. MANIPUR	
Total Geographic Area	22,327 Km^2
Recorded forest area	243 Km^2
Forest area of TGA	1.09 per cent
Annual Rainfall	1881 mm
Total Protected areas	3
National Park	1
Wildlife Sanctuary	2
State animal	Sangai
State bird	Nongyeen
State tree	Unigthou
Local tribes	Aimol, Chiru, Mao, Thadou
17. MEGHALAYA	
Total Geographic Area	22,429 Km^2

Recorded forest area	710 Km^2
Forest area of TGA	3.17 per cent
Annual Rainfall	2818 mm
Total Protected areas	5
National Park	2
Wildlife Sanctuary	3
State animal	Clouded Leopard
State bird	Hill Myna
State tree	White Teak
Local tribes	Garo, Khasi, Jaintia, Boro
18. MIZORAM	
Total Geographic Area	21,081 Km^2
Recorded forest area	535 Km^2
Forest area of TGA	2.54 per cent
Annual Rainfall	1881 mm
Total Protected areas	10
National Park	2
Wildlife Sanctuary	8
State animal	Serow
State bird	Hume's Bartailed Pheasant
State tree	Nahar
Local tribes	Lushai, Chakma, Kuki
19. NAGALAND	
Total Geographic Area	16,579 Km^2
Recorded forest area	381 Km^2
Forest area of TGA	2.30 per cent
Annual Rainfall	1881 mm
Total Protected areas	4
National Park	1
Wildlife Sanctuary	3
State animal	Mithun
State bird	Tragopan Blythu
State tree	Alder
Local tribes	Nagas, Ao, Chang, Lotha, Mikir

20. ODISHA	
Total Geographic Area	1,55,707 Km^2
Recorded forest area	3,986 Km^2
Forest area of TGA	2.56 per cent
Annual Rainfall	1489 mm
Total Protected areas	21
National Park	2
Wildlife Sanctuary	19
State animal	Sambar
State bird	Indian roller
State tree	Aswatha
Local tribes	Khond, Gond, Juang, Baiga, Ho
21. PUNJAB	
Total Geographic Area	50,362 Km^2
Recorded forest area	1,544 Km^2
Forest area of TGA	3.07 per cent
Annual Rainfall	649 mm
Total Protected areas	13
National Park	0
Wildlife Sanctuary	13
State animal	Blackbuck
State bird	Baaz
State tree	Sheesham
Local tribes	Bangali, Bazigar, Garo, Perna
22. RAJASTHAN	
Total Geographic Area	3,42,239 Km^2
Recorded forest area	8,269 Km^2
Forest area of TGA	2.42 per cent
Annual Rainfall	650 mm
Total Protected areas	30
National Park	5
Wildlife Sanctuary	25
State animal	Camel/Chinkara
State bird	Great Indian Bustard

State tree	Khejri
Local tribes	Bhil, Meena, Shariya, Gujjar
23. SIKKIM	
Total Geographic Area	7,096 Km^2
Recorded forest area	35 Km^2
Forest area of TGA	0.50 per cent
Annual Rainfall	2739 mm
Total Protected areas	8
National Park	1
Wildlife Sanctuary	7
State animal	Red Panda
State bird	Blood Pheasant
State tree	Rhodendron
Local tribes	Bhutia, Lepcha, Serpa
24. TAMIL NADU	
Total Geographic Area	1,30,058 Km^2
Recorded forest area	4,505 Km^2
Forest area of TGA	3.46 per cent
Annual Rainfall	998 mm
Total Protected areas	34
National Park	5
Wildlife Sanctuary	29
State animal	Nilgiri Thar
State bird	Emerlad Dove
State tree	Palmyra Palm
Local tribes	Adiyan, Kadar, Toda,
25. TELANGANA	
Total Geographic Area	1,14,865 Km^2
Recorded forest area	2549 Km^2
Forest area of TGA	2.22 per cent
Annual Rainfall	961 mm
Total Protected areas	12
National Park	3
Wildlife Sanctuary	9

State animal	Deer
State bird	Indian roller
State tree	Jammi
Local tribes	Andh, Bhil, Chenchu, Gond
26. TRIPURA	
Total Geographic Area	10,486 Km^2
Recorded forest area	233 Km^2
Forest area of TGA	2.22 per cent
Annual Rainfall	1881 mm
Total Protected areas	6
National Park	2
Wildlife Sanctuary	4
State animal	Phayre's Langur
State bird	Green Imperial Pigeon
State tree	Agar
Local tribes	Chakamas, Orang, Santal, Garoo
27. UTTAR PRADESH	
Total Geographic Area	2,40,928 Km^2
Recorded forest area	7,044 Km^2
Forest area of TGA	2.92 per cent
Annual Rainfall	1250 mm
Total Protected areas	25
National Park	1
Wildlife Sanctuary	24
State animal	Swamp Deer
State bird	Saras Crane
State tree	Ashok
Local tribes	Bhotia, Buska, Raji, Tharu
28. UTTARAKHAND	
Total Geographic Area	53,483 Km^2
Recorded forest area	752 Km^2
Forest area of TGA	1.41 per cent
Annual Rainfall	650 mm
Total Protected areas	13

National Park	6
Wildlife Sanctuary	7
State animal	Musk Deer
State bird	Himalayan Monal
State tree	Burans
Local tribes	Bhotia, Buska, Raji, Tharu
29. WEST BENGAL	
Total Geographic Area	88,752 km^2
Recorded forest area	2,088 km^2
Forest area of TGA	2.35 per cent
Annual Rainfall	2500 mm
Total Protected areas	21
National Park	6
Wildlife Sanctuary	15
State animal	Fishing Cat
State bird	White throated kingfisher
State tree	Chatian
Local tribes	Asur, Gond, Garo, Munda

UNION TERRITORIES FOREST PROFILE

1. ANDAMAN AND NICOBAR ISLANDS	
Total Geographic Area	8,249 Km^2
Recorded forest area	37 Km^2
Forest area of TGA	0.44 per cent
Annual Rainfall	2967 mm
Total Protected areas	102
National Park	9
Wildlife Sanctuary	96
State animal	Sea Cow
State bird	Andman Wood Pigeon
State tree	Andman Paduk
Local tribes	Andamanese, Onges, Bo, Tabo

2. CHANDIGARH	
Total Geographic Area	114 Km^2
Recorded forest area	9 Km^2
Forest area of TGA	7.80 per cent
Annual Rainfall	617 mm
Total Protected areas	2
National Park	0
Wildlife Sanctuary	2
State animal	Indian Grey Mangoose
State bird	Indian Grey Hornbill
State tree	Mango Tree
Local tribes	Baleniki, Khatik, Nat
3. DADRA AND NAGAR HAWELI	
Total Geographic Area	491 Km^2
Recorded forest area	28 Km^2
Forest area of TGA	5.72 per cent
Annual Rainfall	2000-2500 mm
Total Protected areas	1
National Park	0
Wildlife Sanctuary	1
State animal	–
State bird	–
State tree	–
Local tribes	Dhodia, Kathodi, Konka, Koli
4. DAMAN AND DIU	
Total Geographic Area	112 Km^2
Recorded forest area	10 Km^2
Forest area of TGA	9.18 per cent
Annual Rainfall	1899 mm

Total Protected areas	11
National Park	0
Wildlife Sanctuary	11
State animal	–
State bird	–
State tree	–
Local tribes	Dhodia, Dubla, Siddi, Varli
5. DELHI	
Total Geographic Area	1,483 Km^2
Recorded forest area	111 Km^2
Forest area of TGA	7.49 per cent
Annual Rainfall	617 mm
Total Protected areas	1
National Park	0
Wildlife Sanctuary	1
State animal	Nilgai
State bird	House sparrow
State tree	–
Local tribes	–
6. LAKSHADWEEP	
Total Geographic Area	32 Km^2
Recorded forest area	4 Km^2
Forest area of TGA	12.50 per cent
Annual Rainfall	1515 mm
Total Protected areas	1
National Park	0
Wildlife Sanctuary	1

State animal	Butterfly Fish
State bird	Sooty Tern
State tree	Bread Fruit
Local tribes	Aminidivi, Koya, Malmi
7. PUDUCHERRY	
Total Geographic Area	480 Km^2
Recorded forest area	27 Km^2
Forest area of TGA	5.63 per cent
Annual Rainfall	998 mm
Total Protected areas	1
National Park	0
Wildlife Sanctuary	1
State animal	Squirrel
State bird	Asian Koel
State tree	Vilva Tree
Local tribes	Karaikal, Mahe, Yanam

(**Source:** *isfr. 2015*)

List of important forest research institutes

- Advanced Research Centre for Bamboo and Rattan, Aizawl
- Agricultural and Processed Food Products Export Development Authority (APEDA)- New Delhi
- Arid Forest Research Institute (AFRI), Jodhpur
- Botanical Survey of India (BSI), Kolkata
- C.P.R. Environmental Education Centre, Chennai
- Center for Animals and Environment, Bangalore
- Center for Ecological Sciences, Bengaluru
- Center for Environmental Management of Degraded Ecosystem, Delhi
- Center for International Forestry Research (CIFOR), Bogor, Indonesia

- Center for Mining Environment, Dhanbad
- Center for Tropical Forest Science, Panama City, Panama
- Center of Excellence in Environmental Economics, Chennai
- Central Drug Research Institute (CDRI) - Lucknow, U.P
- Central Zoo Authority of India (CZI), New Delhi
- Centre for Environment Education, Ahmedabad
- Centre for Forest-based Livelihoods and Extension (CFLE), Agartala
- Centre for Forestry Research and Human Resource Development, Chhindwara
- Centre for Social Forestry and Eco-Rehabilitation, Allahabad
- Directorate of Forest Education, Dehradun
- European Forest Institute, Joensuu, Finland
- Forest Research Institute (FRI), Dehradun
- Forest Survey of India (FSI), Dehradun
- Foundation for Revitalizations of Local Health Traditions (FRLHT), Bengaluru
- Govind Ballabh Pant Institute of Himalayan Environment & Development, Almora
- Himalayan Forest Research Institute (HFRI), Shimla
- Indian Grassland and fodder research institute (IGFRI)- Jhansi, Uttar Pradesh
- Indian Institute of Forest Management (IIFM), Bhopal
- Indian Institute of Plantation Management (IIPM), Bangalore
- Indian Institute of Soil science (IISS)-Bhopal, MP.
- Indian Lac research institute (ILRI)-Ranchi, Jharkhand
- Indian Plywood Industries Research and Training Institute, Bangalore
- Indira Gandhi National Forest Academy (IGNFA), Dehradun
- Institute of Forest Biodiversity (IFB), Hyderabad
- Institute of Forest Genetics and Tree Breeding (IFGTB), Coimbatore
- Institute of Forest Productivity (IFP), Ranchi
- Institute of Wood Science and Technology (IWST), Bangalore
- International Bureau of Plant Genetic Resources (IBPGR), Rome, Italy
- International Union of Forest Research Organizations (IUFRO) Vienna, Austria

- National Bank for Agriculture and Rural Development (NABARD)-Mumbai
- National Biodiversity Authority (NBI), Chennai
- National Biodiversity Board (NBB), New Delhi
- National Bureau of Plant Genetic Resources (NBPGR), New Delhi
- National Bureau of Soil Survey and Land Use Planning (NBSS & LUP)-Nagpur
- National Ganga River Basin Authority (NGRBA), New Delhi
- National Institute of Animal Welfare (NIAW), Faridabad
- National Museum of Natural History, New Delhi
- National Tiger Conservation Authority (NTCA), New Delhi
- National Zoological Park, New Delhi
- Rain Forest Research Institute (RFRI), Jorhat
- Salim Ali Center for Ornithology and Natural History (SACON), Coimbatore
- The Convention on International Trade in Endangered Species of Wild Fauna and Flora (CITES)-Washington-D
- The International Union for Conservation of Nature and natural resources (IUCN)-Switzerland
- Tropical Botanic Garden and Research Institute (TBGRI), Thiruvananthapuram
- Tropical Forest Research Institute (TFRI), Jabalpur
- Wildlife Crime Control Bureau (WCCB)-New Delhi
- Wildlife Institute of India (WII), Dehradun
- World Agroforestry Center (WAC)- Nairobi, Kenya
- Zoological Survey of India (ZSI), Kolkata
- Per capita forest area available in India is 0.06 hectare.
- Per capita forest area available in world is 0.64 hectare.
- Total volume of water in the hydrosphere is 1.4 billion cubic kilometres, about **97.5%** is ocean water and unsuitable for human use. While **2.5%** is fresh water.
- About 1.97% of fresh water is stored in the form of ice-caps and glaciers.
- Maximum Angiospermic plants are found in Tamil Nadu.
- Maximum Gymnosperm plants are found in Andhra Pradesh.

- 19th Commonwealth Forestry Conference was held on 3rd April 2017 at Forest Research Institute Dehra Dun, India.
- The theme of 2017 **International Day of Forest** is ***"Forest and Energy."***
- The theme of 2017 International Day of Water is ***"Why Waste Water."***
- Chipko movement was started in 1974 at Gopeshwar in Chamoli district by Sunder Lal Bahuguna.
- Apiko movement was started in 1983.
- Indian forest service was started in 1966.
- There are three Cadres in All India Services.

Name of Cadre	Established in
Indian Administrative Service (IAS)	1893
Indian Police Service (IPS)	1948
Indian Forest Service (IFS)	**1966**

- Selvas in Brazil are dense equatorial forest.
- Golden Triangle in South East Asia is known in the world for opium production.
- There is four pressure belt are found on our globe. Out of four two are high pressure belt and the other two are low pressure belt.
- Chilgoza is obtained from *Pinus gerardiana*.
- **Norway** is the **first** country in the World to **ban deforestation**.
- Maharashtra declared **Blue Morman** as state butterfly.
- Average rainfall of India is 1194 mm.
- Annual world precipitation is 1000 mm.
- Geographical area of India is 328.74 million hectare.
- National research agroforestry centre situated in Jhansi, Uttar Pradesh.
- Symbol of World Wide Fund for Nature is Red Panda.
- WWF is now replaced by World Wide Fund for Nature (WFN).
- Methyl Isocynate (MIC) was released in Bhopal gas Tragedy on 3rd December 1984.
- IUCN is now replaced by World Conservation Union (WCU).
- Regur refers to black soil. Most predominant soil type in India is red soil.

- 11th Ministerial Conference of WTO was held in Buenos, Argentina.
- National Seed Research & Training Centre located in Vanaras, Uttar Pradesh.
- The theme of 2018 International Day of Forest is "*Forest and Sustainable Cities*"
- The theme of 2018 International Day of Water is "*Nature for Water.*"
- The theme of 2019 International Day of Forest is "*Forest and Education*"
- The theme of 2019 International Day of Water is "*Life under Water.*"
- COP 21st was held in Paris, France.

Important Environment Days

Name	Date
World Wetlands Day	Feb-02
International Polar Bear Day	Feb-27
World Wildlife Day	Mar-03
International Day of Action for Rivers	Mar-14
World Consumer Rights Day	Mar-15
World Sparrow Day	Mar-20
International Day of Forests	Mar-21
World Planting Day	Mar-21
World Wood Day	Mar-21
World Water Day	Mar-22
World Fish Migration Day	Apr-21
Earth Day	Apr-22
World Migratory Bird Day	Second Saturday in May
Endangered Species Day	May-20
World Biodiversity Day	May-22
World Environment Day	Jun-05
World Oceans Day	Jun-08
World Giraffe Day	Jun-21
World Population Day	Jul-11
International Tiger Day	Jul-29
World Lion Day	Aug-10
World Elephant Day	Aug-12
National Honey Bee Day	Aug-22

Amazon Day	Sep-05
World Clean-up Day 2018	Sep-08
International Ozone day	Sep-16
World Rhino Day	Sep-22
World Environmental Health Day	Sep-26
World Rivers Day	last Sunday in September
World Habitat Day	first Monday in October
World Animal Day	Oct-04
International Day of Climate Action	Oct-24
World Science Day	Nov-01
America Recycles Day	Nov-15
World Soil Day	Dec-05
International Mountain Day	Dec-11

List of important Year

World Population Year	1974
International Year of the Ocean (IYO)	1998
International Year of Mountains (IYM)	2002
International Year of Ecotourism (IYE)	2002
International Year of Freshwater (IYF)	2003
International Year of Deserts and Desertification	2006
International Year of the Dolphin	2007–2008
International Polar Year	2007–2009
International Year of Planet Earth	2008
International Year of Sanitation	2008
International Year of Natural Fibres	2009
Year of the Gorilla	2009
International Year of Biodiversity	2010
International Year of Forests	2011
International Year of Soils	2015
International Year of Pulses	2016

Indian forest Act-1927

Indian forest Act-1927 contains 13 chapters and 86 sections.

Chapter Name	Deals with
Chapter i	Preliminary
Chapter ii	Reserved forests
Chapter iii	Village-forests
Chapter iv	Protected forests
Chapter v	Control over forests and lands not being the property government
Chapter vi	The duty on timber and other forest-produce
Chapter vii	The control of timber and other forest-produce in transit
Chapter viii	The collection of drift and stranded timber
Chapter ix	Penalties and procedure
Chapter x	Cattle-trespass
Chapter xi	Forest-officers
Chapter xii	Subsidiary rules
Chapter xiii	Miscellaneous

Wildlife Protection Act-1972

Contains 7 chapters, 6 schedules and 66 sections.

Chapter Name	Deals with
Chapter i	Preliminary
Chapter ii	Authorities to be appointed or constituted under this Act
Chapter iii	Hunting of Wild Animals
Chapter iv	Sanctuaries, National Parks and Closed Areas
Chapter v	Trade or Commerce in Wild Animals,
Chapter vi	Prevention and Detection of Offences
Chapter vii	Miscellaneous
Schedule	**Deals with**
Schedule i	Conservation of rare and endangered species
Schedule ii	Special game animals
Schedule iii	Big game animals
Schedule iv	Small game animals
Schedule v	Vermin animals
Schedule vi	Cultivation of plants

- Indian Panel Code-1860
- Criminal Procedure code-1973
- Water (Prevention & control of Pollution) Act-1974
- Forest Conservation Act – 1980
- Air (Prevention & control of Pollution) Act-1981
- National Environment Protection Act-1986
- Biosphere Reserve Schemes -1986
- Biodiversity Act- 2002

List of important instruments

Most commonly used instruments list are given below:

Device	Quantity measured
Actinometer	Heating power of sunlight
Altimeter	Altitudes
Anemometer	Wind speed
Audiometer	Hearing
Barometer	Air pressure
Bomb calorimeter	Heat of combustion
Calorimeter	Heat of chemical reactions
Chronometer	Time
Compass	Direction of North
Crescograph	Plant growth
Densimeter	Specific gravity of liquids
Dumpy Level	Horizontal levels
Evaporimeter	Rate of evaporation
Fathometer	Ocean depth
Glucometer	Blood glucose (diabetes)
Hydrometer	Specific gravity of liquids (density of liquids)
Hygrometer	Humidity
Clinometer	Angle of a slope
Lactometer	Specific gravity of milk
Manometer	Pressure of gas
Mercury Barometer	Atmospheric pressure
Odometer	Distance travelled
Osmometer	Osmotic strength of solution

Pedometer	Steps
PH Meter	PH (chemical acidity/basicity of a solution)
Photometer	Light intensity
Potometer	Transpiration
Planometer	Area
Psychrometer	Humidity
Pycnometer	Fluid density
Pyranometer	Solar radiation
Rain Gauge	Measuring of rain
Saccharometer	Amount of sugar in a solution
Seismometer	Seismic waves (for example, earthquakes)
Spectrophotometer	Intensity of light as a function of wavelength
Sphygmomanometer	Blood pressure
Tensiometer	Surface tension of a liquid
Theodolite	Measuring angles in the horizontal and vertical planes
Thermometer	Temperature
Tintometer	Colour
Wind Vane	Wind direction
Zymometer	Fermentation

List of important conferences:

Conference Name	Theme of Conference	Year	Held at
Human Environment	Environment conservation	1972	Stockholm
Montreal Protocol	Ozone protection	1987	Montreal, Canada
Earth Summit	Biodiversity conservation & Reducing GHGs	1992	Rio-de-Jenerio
Kyoto Protocol	Climate change	1997	Kyoto, Japan
United Nation	Desertification	1997	Nairobi, Kenya
World Summit	Sustainable development	2002	Johncsberg

Terms related to Forestry:

1. **Agriculture:** Latin word-“ager” meaning ‘soil’ and “cultura” meaning ‘cultivation’.
2. **Extension:** Latin word-“ex” meaning ‘out’ and “tensio” meaning ‘stretching’.

3. **Soil Science:** Latin word-"catena" meaning 'chain'.
4. **Soil:** Latin word-"solum" meaning 'ground'
5. **Monsoon:** Arabic word –"mausim" meaning 'season'.
6. **Entomology:** Greek word –"entomo" meaning 'insect' and "logos" meaning 'study'.
7. **Ecology:** Greek word –"oikos" meaning 'habitat' and "logos" meaning 'study'.
8. **Meteorology:** Greek word –"meteoro" meaning 'above earth surface' and "logy" meaning 'indicating science'.
9. **Forest mensuration:** Latin word-"Mensura" meaning 'measure'.
10. **Silviculture:** Latin word- "silvi" meaning 'forest and 'culture" meaning 'as in growing'.
11. **Dendrology:** Greek word –"dendron" meaning 'tree' and "logia" meaning 'study'.
12. **Forest Pathology:** Greek word –"pathos" meaning 'ailments' and "logos" meaning 'study'.
13. **Ergonomics:** Greek word –"ergon" meaning 'work' and "nomoi" meaning 'natural law'.
14. **Communication:** Latin word-"communis" meaning 'common'.
15. **Economics:** Greek word –"oikonomia" meaning 'household management'.

List of important revolutions

Name	Related with
Green revolution	Rice and wheat production
White revolution	Milk production
Blue revolution	Fish production
Yellow revolution	Oilseeds production
Red revolution	Tomato/meat production
Golden revolution	Fruit production
Round revolution	Potato production
Black revolution	Biofuel/Jatropa revolution
Pink revolution	Onion production
Rainbow revolution	All sector of agriculture

Agro-climatic zones of India

India has fifteen (15) agro-climatic zones.

Agro-climatic zone	Annual Rainfall (cm)	Occupied region	Agriculture crops
1. Western Himalayan Region	75-150	Jammu and Kashmir, Himachal Pradesh, and the hill region of Uttarakhand.	Crops-Rice, maize Fruit crops-Apple, peaches, apricot, pears, cherry, almond, litchis, walnut, Saffron
2. Eastern Himalayan Region	200-400	Arunachal Pradesh, the hills of Assam, Sikkim, Meghalaya, Nagaland, Manipur, Mizoram, Tripura, and the Darjeeling district of West Bengal	Soil- Red-brown, Jhuming (shifting cultivation) prevails in the hilly areas. Crops- Rice, maize, potato, tea Fruit crops-Pineapple, litchi, oranges and lime.
3. Lower Gangetic Plain Region	100-200	West Bengal (except the hilly areas), eastern Bihar and the Brahmaputra valley	Crops -Rice Jute, maize, potato, and pulses. Fruit crops -Banana, mango and citrus fruits.
4. Middle Gangetic Plain Region	100-200	The Middle Gangetic Plain region includes large parts of Uttar Pradesh and Bihar	Soil- Fertile alluvial. Crops- Rice, maize, millets, wheat, gram, barley, peas, mustard and potato
5. Upper Gangetic Plains Region	75-150	Central and Western parts of Uttar Pradesh and Haridwar and Udham Nagar districts of Uttarakhand	Soil- Sandy loam Crops-Wheat, rice, sugarcane, millets, maize, gram, barley, oilseeds, pulses and cotton

6. Trans-Ganga Plains Region	65-125	Punjab, Haryana, Chandigarh, Delhi,Ganganagar district of Rajasthan	Soil-Alluvial Crops-wheat, sugarcane, cotton, rice, gram, maize, millets, pulses and oil-seeds
7. Eastern Plateau and Hills	80-150	Chhotanagpur Plateau, extending over Jharkhand, Odisha, Chhattisgarh and Dandakaranya	Soils- Red and Yellow soil Crops- Rice, millets, maize, oilseeds, ragi, gram and potato, tur, groundnut, soybean urad, castor
8. Central Plateau and Hills	50-100	Bundelkhand, Baghelkhand, Bhander Plateau, Malwa Plateau, and Vindhyachal Hills.	Soils-Red, yellow and black. Crops-Millets, wheat, gram, oilseeds, cotton and sunflower
9. Western Plateau and Hills	25-75	southern part of Malwa plateau and Deccan plateau (Maharashtra)	Soil- Regur (black) soil Crops-Wheat, gram, millets, cotton, pulses, groundnut, oilseeds, sug-arcane, rice, jowar, bajra Fruit crops- Oranges, grapes, bananas, ber, pomegranate, mango and guava (sprinklers and drip system)
10. Southern Plateau and Hills	50-100	Interior Deccan and includes parts of southern Maha-rashtra, the greater parts of Kar-nataka, Andhra Pradesh, and Tamil Nadu	Crops- Millets, oilseeds, and pulses are grown. Coffee, tea, cardamom and spices

11. Eastern Coastal Plains and Hills	75-150	Coromandal and northern Circar coasts of Andhra Pradesh and Orissa.	Soils -Alluvial, loam and clay Crops-Rice, jute, tobacco, sugarcane, maize, millets, groundnut and oilseeds, pepper and cardamom
12. Western Coastal Plains and Ghats	more than 200	Malabar and Konkan coastal plains and the Sahyadris,	Soils -Laterite and alluvial Crops-Rice, coconut, oilseeds, sugarcane, millets, pulses and cotton
13. Gujarat Plains and Hills	50-100	The hills and plains of Kathiawar, and the fertile valleys of Mahi and Sabarmati rivers.	Soils- Regur & alluvium. Crops-Groundnut, cotton, rice, millets, oilseeds, wheat and tobacco are the main crops. It is an important oilseed producing region.
14. Western Dry Region	less than 25	Extending over Rajasthan, West of the Aravallis,	Crops- Bajra, jowar, moth, wheat and gram Fruit crops-Water melon, guava and date palm,
15. Island Region	less than 300	Andaman-Nicobar and Lakshadweep	Soil- Sandy loam Crops- Rice, maize, millets, pulses, arecanut, turmeric and cassava. Nearly half of the cropped area is under coconut.

(According to Planning commission of India)

Important terminology related to animals

Animal	Young	Female	Male	Group	meat	Avg. gestation period (in days)
Antelope	Calf	Doe	Buck	Herd	-	-
Ape	Infant	-	-	Troop	-	-
Ass/donkey	Foal	Jenny	Jack	Herd	-	365

Bat	Pup	-	-	Cloud	-	-
Bear	Cub	Sow	Boar	Sleuth	-	220
Bee	Larva	Queen	Drone	Swarm	-	-
Bird	Chick	Hen	Cock	Flock	-	-
Bison	Calf	Cow	Bull	Herd	-	217
Boar (wild pig)	Piglet	Sow		Herd	-	-
Buffalo	Calf	Cow	Bull	Herd	Buffin	316
Camel	Calf	Cow	Bull	Caravan		390
Caribou	Calf	Doe	Buck	Herd		-
Cat	Kitten	Molly	Tom	Clowder	-	64
Cattle	Calf	Cow	Bull	Herd	Beef	286
Cheetah	Cub	-	-	Coalition		
Chicken	Chick	Hen	Rooster, cock	Flock	Poultry	-
Chimpanzee	Infant	Empress	Blackback	-		240
Crane	Chick	-	-	Herd		-
Crocodile	Hatchling	Cow	Bull	Bask		
Deer	Fawn	Doe	Buck	Herd	Venison	201
Dinosaur	Hatchling	Cow	Bull	Herd		
Dog	Pup	Bitch	Dog	Pack	-	61
Dolphin	Calf	Cow	Bull	School	-	-
Duck	Duckling	Duck	Drake	On land	Poultry	-
Elephant	Calf	Cow	Bull	Herd	-	617
Elk	Calf	Cow	Bull	Gang	Venison	245
Emu	Chick	-	-	Mob		-
Gaur	Calf	Cow	Bull	Herd	Gara-beef	215
Giant panda	Cub	Sow	Boar	Sleuth	-	-
Giraffe	Calf	Cow	Bull	Herd	-	430
Goat	Kid	Nanny	Billy	Herd	Chevon	150
Gorilla	Infant	-	Blackback	Band	-	257
Hippopotamus	Calf	Cow	Bull	Bloat	-	237
Horse	Foal	Mare	Stallion	Herd	-	336
Humming-bird	Chick	Hen	Cock	-	-	-
Hyena	Cub	Bitch	Dog	Cackle	-	-
Kangaroo	Joey	Flyer	Boomer	Court	-	42

Leopard	Cub	Leopardes	Leopard	Leap	-	93
Lion	Cub	Lioness	Lion	Pride	-	108
Mongoose	Pup	-	-	Troop	-	-
Monkey	Infant	-	-	Troop	-	64
Moose	Calf	Cow	Bull	Herd	Venison	-
Mouse	Kitten	Doe	Buck	Colony	-	19
Mosquito	Nymph	-	-	Swarm	-	-
Nightingale	Chick	Hen	Cock	Flock	-	-
Ostrich	Hatchling	Hen	Cock	Flock	-	-
Panther	Cub	Pantheress	-	-	-	-
Parrot	Chick	Hen	Cock	Flock	-	-
Penguin	Chick	-	-	Waddle	-	-
Pheasant	Chick	Hen	Cock	Flock	-	-
Pig	Piglet	Sow	Boar	Drift	Pork	113
Pigeon	Squab	Hen	Cock	Flock	Squab	-
Polar bear	Cub	Sow	Boar	Aurora	-	-
Porcupine	Pup	Sow	Boar	Prickle	-	210
Rabbit	Bunny	Doe	Buck	Nest	-	31
Sheep	Lamb	Ewe	Ram	Flock	Mutton	147
Rat	Kitten	Doe	Buck	Colony	-	22
Red deer	Calf	Hind	Stag	Herd	Venison	208
Red panda	Cub	Sow	Boar	Pack	-	-
Reindeer	Calf	Cow	Bull	Herd	-	-
Rhinoceros	Calf	Cow	Bull	Herd	-	450
Sea lion	Pup	Cow	Bull	-	-	-
Seahorse	Seafoal	Seamare	Seastallion	Shoal	-	-
Seal	Pup	Cow	Bull	Herd	-	330
Sheep	Lamb	Ewe	Ram	Flock	Lamb	-
Snake	Snakelet	-	-	Nest	-	-
Squirrel	Pup	Doe	Buck	Scurry	-	35
Tiger	Cub	Tigress	Tiger	Ambush	-	109
Walrus	Calf	Cow	Bull	Herd	-	-
Water buffalo	Calf	Cow	Bull	Herd	Carabeef	-
Whale	Calf	Cow	Bull	Pod	Blubber	535
Wolf	Pup	Bitch	Dog	Pack	-	64
Yak	Calf	Cow	Bull	Herd	-	-
Zebra	Colt	Mare	Stallion	Herd	-	375

Important projects related to wildlife and the year of launching

Project Name	Year
Project Hangul	1970
Project Lion	1972
Project Tiger	1973
Crocodile breeding Project	1975
Manipur Bow deer Project	1977
Project Elephant	1991
Project Musk deer	1991
Project Great Indian bustard	1993
Project Siberian crane	1995
Project Snow leopard	2009

Location of national parks in India

State	National Park
Haryana	• Kalesar National Park • Sultanpur National Park
Himachal Pradesh	• Pin Valley National Park • Inderkilla National Park • Great Himalayan National Park • Khirganga National Park • Simbalbara National Park
Jammu & Kashmir	• Dachigam National Park • Salim Ali National Park • Kishtwar National Park • Hemis National Park
Uttar Pradesh	• Dudhwa National Park
Gujarat	• Blackbuck National Park, Velvadar • Gir Forest National Park • Marine National Park, Gulf of Kutch • Bansda National Park
Goa	• Bhagwan Mahavir (Mollem) National Park
Telangana	• Mrugavani National Park
Assam	• Dibru-Saikhowa National Park • Rajiv Gandhi Orang National Park • Kaziranga National Park • Manas National Park • Nameri National Park
Bihar	• Valmiki National Park

Chhattisgarh	• Indravati National Park • Guru Ghasi Das (Sanjay) • Kanger Valley National Park
Jharkhand	• Betla National Park
Andhra Pradesh	• Sri Venkateswara National Park
Madhya Pradesh	• Kanha National Park • Madhav National Park • Van Vihar National Park • Mandla Plant Fossils National Park • Panna National Park • Pench National Park • Sanjay National Park • Satpura National Park
Karnataka	• Bandipur National Park • Anshi National Park • Kudremukh Natiokeynal Park • Bannerghatta National Park • Nagarhole National Park
Kerala	• Pamapadum Shola National Park • Anamudi Shola National Park • Eravikulam National Park • Mathikettan Shola National Park • Silent Valley National Park • Anamudi Shola National Park
Maharashtra	• Chandoli National Park • Gugamal National Park • Tadoba National Park • Navegaon National Park
Meghalaya	• Balphakram National Park • Nokrek National Park
Manipur	• Keibul Lamjao National Park
Mizoram	• Murlen National Park • Mountain National Park • Phawnguni Blue National Park
West Bengal	• Buxa National Park • Sundarbans National Park • Neora Valley National Park • Singalila National Park
Uttarakhand	• Gangotri National Park • Jim Corbett National Park • Valley of Flowers National Park • Rajaji National Park

Rajasthan	• Sariska National Park • Keoladeo National Park • Ranthambore National Park • Darrah National Park • Desert National Park
Nagaland	• Intanki National Park
Odisha	• Bhtarkanika National Park • Simlipal National Park
Tamil Nadu	• Mudumalai National Park • Guindy National Park • Mukurhti National Park • Indira Gandhi (Annamalai) National Park • Gulf of Mannar Marine National Park
Tripura	• Clouded Leopard National Park • Bison (Rajbari) National Park
Sikkim	• Khangchendzonga National Park
Arunachal Pradesh	• Namdapha National Park • Mouling National Park

List of important organizations

Name	Established Year	Headquarter
FRI	1906	Dehra Dun, India
ICAR	1929	New Delhi
RBI	1935	Mumbai, India
AGMARK	1937	Nagpur, Maharashtra
GATT	1948	Geneva, Switzerland
IUCN	1948	Morgis, Switzerland
IMD	1961	Pune, India
WWF	1961	Gland, Switzerland
FCI	1964	New Delhi
WIPO	1967	Geneva, Switzerland
CITES	1975	Washington, D.C.
ICRAF	1977	Nairobi, Kenya
NABARD	1982	Mumbai, India
ITTO	1985	Japan
WTO	1995	Geneva, Switzerland

Botanical name of important trees

Common Name	Botanical name	Family	Other Name
Silver fir	*Abies pindrew*	Pinacea	–
Khair	*Acacia catechu*	Mimoseae	–
Babool	*Acacia nilotica*	Mimoseae	–
Dhonkra	*Anogeissus pendula*	Combretaceae	–
Safed donkra	*Anogeissus latifolia*	Combretaceae	Axle wood
Mahaneem	*Ailanthus excelsa*	Simarubiaceae	Tree of Heaven
Haldu	*Adina cordifoia*	Rubiaceae	–
Pink cedar	*Acrocarpus fraxinifolius*	Legumniosae	–
Aus. Wattle	*Acacia auriculoformis*	Mimoseae	–
Kadamb	*Anthocephalus cinensis*	Rubiaceae	–
Devil tree	*Alostonia scholaris*	Legumniosae	Devil's Tree
Ashok	*Ashoka indica*	Legumniosae	–
Bel	*Aegle marmelos*	Rutaceae	God Fruit
Neem	*Azadirachta indica*	Meliaceae	–
Bamboo	*Bambusa bamboos*	Poaceae	Green Gold
Semal	*Bombax ceiba*	Bombacaceae	Kapok
Charoli	*Buchanania lanzan*	Caesalpinaceae	Chironzi
Kachnar	*Bauhania variegata*	Caesalpinaceae	–
Dhak	*Butea monosperma*	Legumniosae	Forest flame
Amaltas	*Cassia fistlua*	Caesalpinacea	Indian laburnum
Deodar	*Cedrus deodara*	Pineaceae	–
Bottle brush	*Callistimon viminalis*	Myrtaceae	–
Saffron	*Crocus sativus*	Iridaceae	–
Cupress	*Cupresses torulosa*	Cupreseaceae	–
Indian kopak	*Ceiba patendra*	Bombaceae	White Semal
Casuarina	*Casuarina equisitifolia*	Casuarinaceae	Beef Wood
Kakooni	*Cochlosprmum religiosum*	Bixaceae	Silk cotton tree
Kadaya	*Caryota urens*	Arecaceae	Fishtail palm
Cycas	*Cycus revoluta*	Cycadaceae	Sago palm
Rose wood	*Dalbergia latifolia*	Paplionaceae	Rose wood
Shisham	*Dalbergia sisso*	Paplionaceae	Biradi
Solid bamboo	*Dendrocalamus strictus*	Poaceae	Male bamboo
Gurjan	*Dipterocarpus turbinatus*	Dipterocarpaceae	–
Gulmohar	*Delonix regia*	Malvaceae	–

Tendu	*Diospyrus melanoxylon*	Ebenaceae	Bedi leaf
Safeda	*Eucalyptus globulus*	Myrtaceae	–
Mysoore gum	*Eucalyptus tereticornis*	Myrtaceae	–
Aonla	*Emblica officinalis*	Euphorbiaceae	–
Eugenia	*Eugenia oojensis*	Myrtaceae	–
Malawar wood	*Hopea parviflora*	Dipterocarpaceae	Malawar wood
Kanju	*Holoptelia integrifolia*	Ulmaceae	Indian elm
Kurchi	*Holorhhinea antidycentrica*	Apocynaceae	Kutuj
Bergad	*Ficus benghalensis*	Moraceae	–
Gular	*Ficus glomeruta*	Moraceae	–
Pepal	*Ficus religiosa*	Moraceae	–
Ghamar	*Gmelia arborea*	Lamiaceae	White teak
Gutel	*Grewia tilifolia*	Malvaceae	Dhaman
Ratanjot	*Jatropha curcus*	Euphorbiaceae	Biodiesal plant
Walnut	*Juglus regia*	Juglandaceae	–
Lendi crape	*Lagerstomia parviflora*	Lytraceae	–
Subabool	*Leucaena leucocephala*	Mimoseae	–
Kaith	*Limonia acidica*	Rutaceae	–
Mahua	*Madhuca indica*	Sapotaceae	Butter tree
Sehtoot	*Morus alba*	Moraceae	Mulberry
Aam	*Mangefera indica*	Anacardaceae	–
Bakain	*Melia azadirache*	Meliaceae	–
Khirni	*Manilkara hexandra*	Sapotaceae	–
Sehjana	*Moringa oleifera*	Moringaceae	Drumstick tree
Indian trumpet	*Oroxylum indicum*	Bignoniaceae	Midnight horror
Kesiya pine	*Pinus kassiya*	Pineaceae	–
Chilgoza pine	*Pinus gerardiana*	Pineaceae	–
Chir pine	*Pinus roxburghii*	Pineaceae	–
Blue pine	*Pinus wallichiana*	Pineaceae	–
Bijasal	*Pterocapus marsupium*	Dipterocaarpaceae	Bengal Kino
Karanj	*Pongamia pinnata*	Leuminosae	–
Spruce	*Picea smithiana*	Pineaceae	–
Popular	*Populus ciliata*	Salicaceae	Himalayan poplar
Chinar	*Platinum orientalis*	Platanaceae	–
Popular	*Populus deltoides*	Salicaceae	–

Black popular	*Populus nigra*	Salicaceae	–
Khejri	*Prosopis cineraria*	Leuminosae	–
Almond	*Prunus amygdalus*	Rosaceae	–
Khejra	*Prosopis juliflora*	Leuminosae	–
Apricoat	*Prunus armoniaca*	Rosaceae	Khubani
Oak	*Quercus rubra*	Fagaceae	Oak tree
Rubinia	*Robinia pseudoacacia*	Fabaceae	–
Chandan	*Santalum alba*	Santalaceae	Sandal wood
Sterculia	*Sterculia urens*	Sterculaceae	–
Sal	*Shorea robusta*	Dipterocaarpaceae	–
Kusum	*Schleria oleosa*	Sapindaceae	Ceylon oak
Aritha	*Sapindus emarginatus*	Euphorbiaceae	–
Salix	Salix alba	Salicaceae	White willow
Sequoia	*Sequoia semperviranes*	Pineaceae	–
Jamun	*Szygigium cuminii*	Myrtaceae	–
Teak	*Tectona grandis*	Verbenaceae	Queen wood
Marwar teak	*Tecomella undulata*	Bignoniaceae	Marwar teak
Arjun	*Terminalia arjuna*	Combretaceae	–
Baheda	*Terminalia bellerica*	Combretaceae	–
Harad	*Terminalia chebula*	Combretaceae	–
Toon	*Toona ciliata*	Meliaceae	Red cedar
Indian laurel	*Terminalia tomentosa*	Combretaceae	–
Kindal tree	*Terminalia peniculata*	Combretaceae	–
Imli	*Tamarindus indica*	Caesalpiniaceae	–
Elm	*Ulmus species*	Ulmaceae	–
Dhudhi	*Wrightia tinctoria*	Apocynaceae	Pala indigo plant
Ber	*Zizyhus mauritiana*	Rhamnaceae	Poor man's apple

List of biosphere reserves in India

Name	Location
1. Achanakmar-Amarkantak	Madhya Pradesh
2. Agasthyamalai	Kerala
3. Cold desert	Laddakh
4. Dehang-Dibang	Arunachal Pradesh
5. Dibru-Saikhowa (smallest)	Assam
6. Great Nicobar	Andaman & Nicobar
7. Gulf of Mannar	Tamil Nadu
8. Kachchh	Gujarat

9. Khangchendzonga	Sikkim
10. Manas (Third)	Assam
11. Nanda Devi (Second)	Uttarakhand
12. Nilgiri (First)	Tamil Nadu
13. Nokrek	Meghalaya
14. Panchmarhi	Madhya Pradesh
15. Seeshachalam Hills	Andhra Pradesh
16. Simlipal	Odisha
17. Sunderbans	West Bengal
18. Panna	Madhya Pradesh

- World 4th Agroforestry Conference in 2019 was held in France.
- Father of Dendrology is Douglas.
- National seed policy-2002.
- Kisan credit card scheme-1998.
- *Dendrocalamus asper* is edible bamboo.
- Bee plan is associated to save elephant (Railway track cross death).
- National green tribunal (NGT) act was formed in 2010.
- Archethobium is smallest parasite.
- National Agroforestry policy was formed in 2014.
- UNFCCC- United Nations Framework Convention on Climate Change.
- Total growing stock of wood in India (FSI, 2017) is 5,822.38 million cubic meter.

CHAPTER 2

Ecology

- Term ecology was given by **E. Haeckel**.
- Father of ecology – **Reiter**.
- **Ecology** – The study of interaction and inter-relationship of organism with their environment.
- Study of the relationship of single species with its environment called **autecology**.
- Study of the relationship of the group of the different species with their environment called **synecology**.
- **Organism** is the smallest unit of ecological hierarchy and basic unit of ecological study.
- **Species** is the basic unit of classification defined as the group of living organism similar in structure, function and can interbreed under natural conditions and reproductively isolated from the other group of organism.
- A species which is found at a particular area is called **endemic species** or endemism.
- The species which have great influence on the community's characteristics relative to their low abundance and biomass are called **key stone species**. The activities of key stone species determine the structure of the community.
- The behaviour of an individual to increase the chances of the survival of other individuals of same species called **altruism**.
- Several members of a species may cover a definite area for searching food, mates and shelter, this is called **home range**.
- In home range particular geographic area/space marked by an individual or group of breeding individuals for breeding called **territory**.
- According to need of light a deep lake has three types of differentiated zone.

1. **Littoral zone:** This zone is found at bank of lake and have maximum diversity. Rooted vegetation is found in this zone.
2. **Limnetic zone:** This zone receive sufficient amount of light on entire surface area. In this zone different types of floating plants, suspended and submerged plants are present.
3. **Profundal zone:** It is the deep area of lake where light does not reach up to the bottom. Only heterotrophs are present in this zone.

- Planktons are free floating small organisms that swim due to water current. They lack locomotory organs or locomotory organs may be reduced.
- Microscopic, inactive floating plants are called **Phytoplankton**.
- Microscopic, inactive floating animals are called **zooplankton**.
- Those aquatic plants or animals which are capable of swimming actively are called **nektons**.
- The sedentary organisms of sea are called **benthonic**.
- Winter sleep or period of dormancy is called **hibernation**. Hibernation is shown by polar bear.
- Summer sleep or escape from heat of sun is called **aestivation**.
- Transition form between two ecotypes is called **Eco-cline**.
- Organism of same trophic level are called guild. Such as cow, goat, rabbit and deer etc.
- Under unfavourable conditions many zooplankton species in lakes and ponds are known to enter **diapause**, a stage of suspended development.
- The phenotypes that show variations due to the difference in the environmental conditions within the local habitat, such types of variations are known as **phenotypic plasticity**.
- Low lying areas covered with shallow water are called **wet lands**. The wet lands are the transitional zones between terrestrial and aquatic areas.
- Member of same species living in different climatic conditions have some genetic variations called **ecotypes** or **ecological races** or **breed**. Genetically as well as morphologically they are dissimilar and interbreeding may be possible.
- Member of same species living in different climatic conditions have some external variations called **ecades or ecophenes**. Genetically they are similar.

- Group of different species that live in common area which are interrelated and interdependent is called a **community**.
- Lianas, epiphytes and epizones show **commensalism**.
- Association in between two organism, when one behave as master and another as a slave is called **helotism**. Lichen shows helotism, in lichen fungi behave as master while algae as slave.
- Association in between the algae and fungi is called **lichen**. While association of fungi with the roots of higher plants is called mychorrhiza.
- Development of plant community on barren area is called **ecological succession or biotic succession.**
- The first community to inhibit an area during the succession called **pioneer community**.
- The last and stable community in an area during the succession called is **climax community**.
- In succession communities or stage which comes in between pioneer community and climax is called **transitional or seral communities**.
- The entire series of communities during succession is called sere.

Succession in fresh water	**Hydrosere**
Succession in salty water	**Halosere**
Succession in acidic water	**Oxalosere**
Succession in dry region	**Xerosere**
Succession on rock	**Lithosere**
Succession in on sand	**Psamosre**
Succession at moistened region	**Mesosere**
Succession of microbes on decomposed matter	**Serula**

- **Primary succession** occurs in the barren area where there is no **previously occuring living** matter present. Afforestation is also a type of primary succession. It takes more time as compare to secondary succession.
- **Secondary succession** occurs where previously vegetation was present but lost due to natural or artificial causes. Reforestation is also a type of secondary succession.
- Savana grassland is a type of **retrogressive succession** (Forest to grass). It means succession occurs in reverse direction.

- **Positive or beneficial interaction**- Member of one or both the interacting species are benefitted but neither is harmed.
- **Negative interaction**- Member of one or both the interacting species are harmed. And there is no obligatory relationship found between interacting species.
- Table showing important association between the organisms. These association may be beneficial or harmful for one or both.

Interaction Types	Beneficial	Harmful	Relationship
Mutualism (+,+)	Both	-	Symbiosis/obligatory
Commensalism (+,0)	One	unaffected	Non-obligatory
Protoco-opration (+,+)	Both	-	**Non-obligatory**
Parasitism (+,-)	One	One (Larger)	Non-obligatory
Predation (+,-)	One	One (smaller)	Non-obligatory
Amensalism (0,-)	unaffected	One	Non-obligatory
Allelopathy (0,-)	unaffected	One	Non-obligatory

- Total root parasite- Rafflesia (Orobanchae).
- Total stem parasite- Cuscuta.
- Partial root parasite- Santalum.
- Partial stem parasite- Viscum (on Oak) and Laranhus (on Mango).
- Organism eaten by their own species is called **cannabilism.**
- Transition zone between two communities is called **ecotone**. It has greater number of species and diversity.
- **Edge effect**- Species which occur most abundantly and spend their time in ecotone are called edge species. The tendency to increase variety and density of some organism at the community border is known as edge effect.
- **Ecological Niche**- Word is given by Grinnel. It is the functional role of any species in an ecosystem or community or it is a functional position or status in an ecosystem.
- **Habitat**- Physical area where an organism lives is called habitat.
- **Gause's competitive exclusion principle**- This principle state that two closely related species competing for the same resources cannot co-exist indefinitely and the competitively inferior will be eliminated eventually. This may be true if resources are limiting but not otherwise.

- **Micro climate and Micro habitat**- Subdivision of habitat is called microclimate. It has an immediate climate (real climate) of an organism which is different from the average climate of that region.
- Term ecosystem was given by **A. G. Tansley**.
- Father of ecosystem ecology is **E. P. Odum**.
- **Ecosystem** is the smallest structural and functional unit of the nature or environment. It is self-regulatory and self-sustaining unit.
- Every ecosystem is composed of two component i.e., **Biotic** or living component and another is **Abiotic** or non-living component.

Biotic components	Abiotic components
Plants	Light
Animals	Temperature
Microbes	Air
	Water
	Soil

- All the autotrophs of an ecosystem called **producer or transducer** or **converter** because they convert solar energy into chemical energy.
- **Solar radiation** is only ultimate source of energy on the earth. Energy enters into the ecosystem through the producers (green plants).
- **Primary consumer** obtained food directly from producers or plants. These all are herbivores of ecosystem. Primary consumer are also called **secondary producers**.
- **Secondary consumers** feed upon primary consumers and obtained their food. Secondary consumers are also called **primary carnivores**.
- Those animals which kill other animal and eat them, but they are not killed and eaten by other animal in nature are called **top consumers.**
- The main **decomposers** in an ecosystem are bacteria, fungi and earthworm.
- In the process of **decomposition** some nutrients get tied up with the biomass of microbes and become temporarily unavailable to other organism, such incorporation of nutrient in living microbes (bacteria, fungi) is called **nutrient immobilization**.
- **Plant parasites** are called primary consumers while animal's parasites are called secondary consumers.

- In an ecosystem every organism depends on other organism for food material and all organism are (herbivore to carnivore) arranged in a series in which food energy is transferred through repeated eating and being eaten. It is called **food chain**. In food chain energy is flow in the form of food.
- Four **trophic level** are present in the ecosystem, because level of energy decreases during the flow of energy from one trophic level to the other trophic level.

First trophic level	[T_1]	= Producers
Second trophic level	[T_2]	= Primary consumer
Third trophic level	[T_3]	= Secondary consumers
Fourth trophic level	[T_4]	= Top consumers

- In food chain energy flow is unidirectional i.e., producer to consumer.
- There are three types of food chains present in nature.
  1. **Grazing food chain or Predatory food chain:** Most of food chain in nature are of this type. Grazing food chain or Predatory food chain begins with producers and in successive order it goes from small organism to big organism.
  2. **Parasitic food chain:** This food chain also starts from producers but in successive order it goes from big organism to small organism.
  3. **Detritus or saprophytic food chain:** This food chain begins with decomposition of dead organic matter by decomposer so it is also called saprophytic food chain.
- Graphical representation of ecological parameters at different trophic level in ecosystem is called **pyramids**. These parameters are Number, Biomass and Energy. First of all pyramid was formed by Charls Elton.
- Pyramid of numbers are mostly upright such as grassland and aquatic ecosystem. But in case of aquatic ecosystem the pyramid of biomass is **inverted**.
- Pyramid of energy are always **upright**.
- According to the **10% law of Lindeman** the 90% part of obtained energy of each organism is utilized in their various metabolic activities and heat and only **10%** energy is transferred to the next trophic level.
- Standing crop- Total amount of living organic matter present in particular area in particular time in an ecosystem is known as **standing crop**.

- Standing state- Total amount of inorganic matter present in a particular area in a particular time in an ecosystem is known as **standing state**.
- Biomass is calculated by the formula **Biomass= Volume x sp.gravity**.
- Pyramid of biomass are mostly upright such as tree ecosystem and forest ecosystem. But in case of tree ecosystem the pyramid of number is inverted.
- **Network** of many food chains is called a food web. The more a food web is complex more stable or permanent the ecosystem is.
- **Productivity**- The rate of biomass production is called productivity. It is expressed in term of **g/m²/yr**.
- **Primary productivity-** The amount of biomass or organic matter produced per unit area over a time period by plants during photosynthesis. It can be divided into GPP and NPP.
- **Gross primary productivity**- The total amount of energy fixed in an ecosystem in unit time is called GPP including the organic matter used up in respiration during the measurement.
- **Net primary productivity-** It is the amount of stored organic matter in plants tissue after respiratory utilization.

  NPP = GPP – R (R- Respiration + Metabolic activity)
- **Secondary productivity**- The rate of formation of new organic matter by consumers.
- **Maximum productivity** per unit area is found in tropical rain forest.
- **Solar radiation** before entering the atmosphere carries energy at a constant rate i.e., **2 Cal cm^{-2} min^{-1}** called solar constant.
- Range of photosynthetically active radiation (PAR) is **400-700 nm**.
- The tails, ears, limbs, snout of mammals are smaller in colder region and larger in warmer region. This is called **Allens rule**.
- The ability of a surface to reflect the incoming radiation is called albido. It is 80% for snow, 20-30% for sand, **5-10% for the forest**. It is 33% for agriculture crops.
- Every organism has minimum and maximum limit of tolerance with respect to environmental factors like temperature, sunlight or nutrient concentration in between those limits the central optimum range are found in which organisms are found in abundance this is known as optimum zone of tolerance. This law called is **Shelford law of tolerance**.

- Megatherms- The plants growing in high temperature throughout the year.
- Mesotherms - The plants growing in alternate high and low temperature.
- Microtherms- The plants growing in low temperature throughout the year.
- Hekistotherms- The plants growing in very low temperature throughout the year.
- The region where river enter into the ocean are known as estuaries.
- The deeper part of ocean where light do not reach is called abyssal zone.
- Height of a place from mean sea level is called altitude while distance of any place from the equator is called latitude.
- Atmosphere may be defined as a transparent gaseous envelope surrounding the earth.

Layer name	Height from earth surface (in Km.)	Key points
Troposphere	0-16	Winds, clouds, dust particles, spores, pollen & more than 90% atmospheric **gases** present.
Stratosphere	16-50	Region for air travel, **ozone layer**
Mesosphere	50-80	-
Thermosphere	80-150	Ionosphere, region of **satellite** launching
Exosphere	Above 150	-

- In troposphere temperature decrease with increase in altitude. The vertical temperature gradient over earth's surface is called lapse rate. It is **6.5°C per 1000 meter**.
- All the living and non-living things of the earth combine together to constitute a big ecosystem called **biosphere**. Biosphere is also called ecosphere.

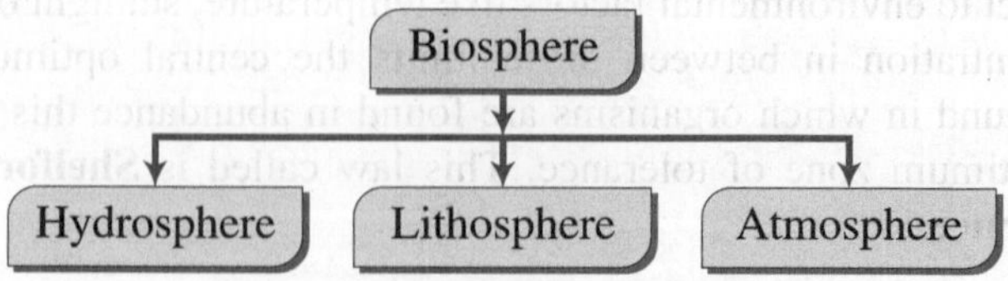

- Smoke plus fog collectively is called smog. This word was given by Desvoeux. Smog is measured by **Ringlmann method.**
- Table showing list of important smog and their formation and final products formed due to their reaction **are given in the table.**

Name of smog	First observed in	Required conditions	Formed product
Los Angeles Smog or Photo Chemical Smog	Los Angeles	Smoke, fog, NO, hydrocarbon.O_2, UV light, and **high** temperature essential.	Reddish brown smog (PAN + O_3 + NO)
London smog or sulphur smog	London	Coal, smoke, fog, SO and **low** temperature	H_2SO_4 vapours

- Due to inhalation of H_2SO_4 vapours with fog 4000 people died in London in **1952.**
- **Acid rain**- This word was given by Robert August. NO_2 and SO_2 released from different sources in the form of smoke and dissolved in atmospheric water vapour to form sulphuric acid and nitric acid. These acids come down on earth with rain water. This is called acid rain.
- The **pH** of acid rain water is less than **5.6.**
- Usually carbon dioxide is not consider as a pollutant but its higher concentration forms a thick layer above the earth's surface and checks the radiation of the heat produced from the earth surface. Because of this, temperature of earth's surface increase this is called **green-house effect** or **global warming.**
- Main green-house gases are **CO_2, CH_4, CFC, N_2O** excluding this SO_2, NO_2, O_3, water vapour are released from industries and agriculture which are responsible to increase the green-house effect.

Green-house gas	Contribution in percent
CO_2	60
CH_4	20
CFC	14
N_2O	06

- At normal temperature and pressure thickness of ozone layer is **3 mm.** The aerosols like chloro floro carbon (CFC) release into atmosphere from the refrigerators, air conditioners and jet planes deplete or reduce the ozone layer. This is called ozone depletion and these substances are called ozone depleting substance. This thin ozone is also known as ozone holes.

- Ozone hole was first discovered by **Nimbus-7** satellite in **1985** over Antarctica.
- Maximum ozone depleting substance is CFCs (**14**) due to release of chlorine. In this process of one chlorine atom convert one lakh O_3 molecules into O_2 by photo dissociation.

$$O_3 \xrightarrow[\text{C.F.C}]{\text{U.V radiation}} \text{oxygen } (O_2) + \text{native oxygen (O)}$$

- Thickness of ozone layer is measured in **Dobson unit (Du)**.
- The amount of dissolved oxygen (D.O) needed by bacteria in decomposing the organic wastes present in water is called **biochemical oxygen demand (B.O.D).**
- Water having D.O content below **8.0 mgL^{-1}** may be considers as contaminated and below **4.0 mgL^{-1}** heavily polluted.
- B.O.D for fresh is less than **1.0 mgL^{-1}.**
- **Daphnia** is the indicator of B.O.D.
- D.O is measured by **oximeter.**
- The oxygen requirement by chemical for oxidation of total organic matter (biodegradable and non-biodegradable) in water is called chemical oxygen demand (C.O.D). C.O.D value is always higher than B.O.D value.
- The non-biodegradable pollutant such as DDT, Al, Hg, Fe etc. are not decomposed by micro-organism. They get accumulated in tissue in increasing concentration along the food chain this is called **biological magnification.**
- The highest concentration of non-biodegradable pollutant occurs in top consumer.
- High concentration of D.D.T. disturb calcium metabolism in birds, which cause thinning of egg shell and their premature breaking, eventually decline in bird population.
- D.D.T. (**Dichloro Diphenyl Tri-chloroethane**) and B.H.C. (Benzene Hexa Chloride) are non-degradable pollutants.
- The process of nutrient enrichment of water and consequent loss of species diversity (or death of aquatic animals) is referred to as

eutrophication and the lake is called eutrophic lake. Its main reason is algal bloom. B.O.D of eutrophic lake is very high.

- Table showing various types of lakes on the basis of accumulation of organic matter in the lake.

Eutrophic lake	Oligotrophic lake	Dystrophic lake
High amount of organic materials and little amount of oxygen. Ex- Dal lake of Kashmir	Less amount of organic materials and little more amount of oxygen.	Maximum amount of undecomposed organic matter. Ex- Marshy lake

- **ELNino effect**- It is the process in which water of Pacific Ocean get warm, in this process warm water current flows to equador and Peru in between 5 to 8 year Christmas time. Effect of **ELNino** is flood, drought and monsoon damaged in India. On the other hand when cold water comes in effect Pacific Ocean it is called **La-Nina effect**.

Name of grassland	Place
Prairies	North America
Pampas	South America
Steppes	Europe and Asia (Russia)
Tussocks	New Zealand
Veldts	Africa
Down	Australia
Savanah	Africa

CHAPTER 3

Forest Management

- According to BCFT Forest management is defined as, 'the practical application of the scientific, technical and economic principles of forestry'.
- Management of forest broadly involves three main tasks viz.,
    1. Control of composition and structure of the growing stock.
    2. Harvesting and marketing of forest produce.
    3. Administration of forest property personnel.
- **95.8%** of Indian forests are under the state ownership.
- India's first forest policy was enunciated in **1894**, which laid down **public benefit** as the sole objective of forest management of public forest.
- Indian Republic formulated its first National Forest Policy in **1952**.
- National Forest Policy in **1952** classified the India's forest into four **classes**

    A. Protection forest   B. National Forest

    C. Village forest   D. Tree-Lands
- National Forest Policy **1952** suggested to keep a minimum of one third of the country's total land area under forests, with 60% in the Himalayas and 20% in the plains.
- The parliament by the 42 Amendment to the Indian Constitution in **1976**, brought Forests and Wildlife on the Concurrent List in the Seventh Schedule.
- **Working plan** is a written scheme of management aiming at continuity of policy, controlling the treatment of a forest. It is based on the principle of sustained yield. It is the unit of forest management.
- **Working Circle**- A forest area organized with particular objects, and under one Silviculture System and one set of Working Plan prescriptions.

Name of classification	Forest classified as
Geographical and climatic (ecological) Classification	• Tropical Rain Forests • Montane Sub Tropical Forests • Montane Temperate • Sub-Alpine • Alpine Scrub
Functional classification	• Protection Forests • National Forests • Village Forests • Tree-Lands
Legal classification	• Reserved Forests • Protected Forests • Village Forests • Un-classed Forests
Territorial classification	• Block • Compartments • Sub-Compartments
Management (Silvicultural) classification	• Working Circles • Felling Series • Cutting Sections • Coupes • Periodic Blocks

- **Block-** It is a main territorial division of forest having a boundary and bearing a local proper name.
- **Compartment-** The smallest permanent Working plan unit of management for administration purposes is called as a compartment. The size of compartment depends on the intensity of management. Compartment is sub division of a block.
- **Sub-Compartment**- Sub division of compartment designated with small letters for temporary administration purpose.
- Head quarter of Inspector General of Forest is situated in **New Delhi**.

Administrative unit	Officer Incharge
State Forest Department	P.C.C.F/ C.C.F
Circle	C.F
Forest Division	D.F.O (D.C.F)

Forest Sub-Division	A.C.F
Range	R.F.O
Section/Block	Forester
Beat	Beat Officer (Forest Guard)

- In our Indian set-up, direct recruitment to the Gazetted rank is made at the level of an **Assistant Conservator of Forest** (A.C.F).
- **Felling Series-** A forest area forming the whole or a part of a Working circle for distribute felling and regeneration to maintain or create a normal distribution of age-classes.
- **Coupe-** A felling area is usually one of an annual series designated with Roman numerals.

  Area of coupe = Area of felling series/Rotation period.
- **Cutting section-** Sub division of felling series formed with the object of regulating cuttings in some special manner.
- **Periodic block-** The part or parts of forest set aside to be regenerated, or otherwise treated during a specified period. The regeneration block is called **floating** or **single** when it is the only periodic block allotted at each Working Plan revision.
- When all periodic blocks are allotted and retain their territorial identity at Working Plan revision, they are called **fixed** or **permanent** periodic blocks.
- The period required to regenerate the whole of a periodic block is called **regeneration period**.
- **Felling cycle-** The time that elapses between successive main fellings on the same area.
- **Age-**Gradation- An age class with one year as the interval.
- The material or cash return obtained from time to time from a forest not organised for continuous production is called **Intermediate Yield**.
- **Growing stock-** The sum (by number or volume) of all the trees growing in the forest or a specified part of it.
- The concept of Progressive Yield was given by German forester **Hartig**.
- The principle of Progressive Yield as against the Sustained Yield principle was discussed at the **sixth Indian Silviculture Conference** held in **Dehra Dun** in 1939 and **third World Forestry Conference** held in **Helsinki** in 1948 and adopted.

- The period which a forest crop takes between its formation and final felling is known as **Rotation or Production period**.
- Rotation which coincides with the natural lease of life of a species on a given site is called **Physical rotation.**
- **Physical rotation** applicable in case of protection and amenity forest, park lands, roadside avenues. This rotation is not of any relevance to economic forestry.
- Rotation through which a species remains satisfactory vigour of growth and reproduction on a given site is called **Silviculture rotation**.
- Silviculture rotation may be useful in forests managed primarily for **aesthetic and recreational purpose**.
- Rotation under which a species yields the maximum material of a specified size or suitability for economic conversion or for special use is called **Technical rotation.**
- Technical rotation is adopted, particularly, by wood based industry such as paper-wood, match-wood industry etc. for maximum raw material's production.
- Rotation that yields the maximum annual quantity of material; i.e., the age at which Mean Annual Increment culminates, is called **rotation of maximum volume production**.
- Rotation of maximum volume production is particularly suitable for adoption where the total quantity of woody material is important and not the **size and specification**.
- Rotation which yields the highest average annual gross or net revenue irrespective of the capital value of the forest is called **Rotation of highest income** or **Rotation of highest revenue** or **Rotation of highest forest rental**. This rotation is important from all national point of view.
- Rotation which yields the highest net return on the invested capital is called **financial or economic rotation**. It differs from the Rotation of highest income in that all items of revenue and expenditure are calculated with compound interest at an assumed rate.
- According to **Hiley** financial or economic rotation is defined as, "the rotation which is most profitable."
- There are two prominent methods to determine the **financial rotation**:
  1. Based on Soil Expectation Value (S_e) of the land.
  2. Based on the financial yield.

- If a piece of land is expected to provide a continual net income of X rupees yearly, then that land can be valued at a sum, which at an acceptable rate of interest gives the same yearly income of Rs. X; that value is known as **Soil Expectation Value (S_e)**.

  $$\text{Soil Expectation Value } (S_e) = \frac{X}{0.0p}$$

  Where **p-rate of interest, percent.**
- The **length of rotation** depends on following parameters-
  1. Rate of growth of species
  2. Silvicultural characteristics of the species
  3. Characteristics of soil
  4. Economic considerations
  5. Social conditions
- The period during which a change from one Silvicultural system to another Silvicultural system is effected is called **conversion period**.
- Conversion period is **usually less** than the rotation period. But it may be sometimes even more or equal to the rotation period.
- Conversion period is usually kept **less** than the rotation period when mature crop is desirable to remove earlier than its rotation period.
- A forest which has normal increment, normal distribution of age-gradation and normal growing forest (NGS) is called **normal forest**.
- **Trinity** of a normal forest is:
  a. Normal Increment.
  b. Normal Distribution of Age-Gradation.
  c. Normal Growing Stock (NGS).
- **Normal Increment** is the best or maximum increment attainable by a given species and for a given rotation, per unit area on a given site.
- **Normal Growing Stock** (NGS) is the volume of a stands in a forest with normal age-classes and a normal increment.
- Kinds of **abnormality** in a forest are given below:
  1. They may be over stocked
  2. They may be under-stocked
  3. They may have normal growing stock volume but abnormal distribution of age classes or age-gradation.

4. The increment may be **sub-normal**
5. Normal increment volume in an **abnormal forest**.

- The increase in girth, diameter, basal area, height, volume, quality and price of individual trees or crops during a given period is called **Increment**.
- The increment in growth that takes place in a particular year is called **Current Annual Increment (CAI)**.
- The increment in growth that takes place for any short period is called **Periodic Annual Increment (PAI)**.
- The total increment up to a given age divided by that age is called **Mean Annual Increment (MAI)**.
- The Mean Annual Increment (MAI) at rotation age called Final Mean Annual Increment.
- The average annual growth in volume over a specified period expressed as a percentage of the volume either at beginning or, more usually, half way through the period is called **Increment Percent**.
- In the beginning **M.A.I** keeps below the **C.A.I**.
- The C.A.I attains the **maximum** before the M.A.I.
- **Current Annual Increment (CAI)** increase very sharp in the beginning but after sometimes it attains its maximum and after this C.A.I. falls to zero but M.A.I never falls up to zero in the whole life of a tree.
- When the C.A.I is equal to M.A.I, this age is the rotation of **maximum volume production**.
- The relation of the increment during a given year to the volume at the beginning of the year is called **Current Increment Percent**.
- The relation of the increment during a given year to a basic volume which may be taken as the mean or average volume for the period, or the volume at the beginning of the period is called **Periodic Increment Percent**.
- The percent ratio which the M.A.I for a given age bears to the total volume at that age is called **Mean Annual Increment Percent**.
- Increment of individual tree can be founded by the Stem-Analysis, by the means of a **Pressler's Increment Borer**, by repeated direct measurement or is taken from the Yield Tables.

- **Pressler's formula** for Increment Percent is given below:

$$\text{Increment Percent } (p) = \frac{V - v}{V + v} \times \frac{200}{n}$$

Where, V = present volume of the crop.

v = Volume of crops is 'n' years ago.

n = Numbers of years (not more than ten years)

- The most convenient formula for finding the increment per cent (p) of standing trees is that developed by **Professor Schneider** in **1853**. This formula is based on the determination of the diameter at breast height, and the number of the rings in the last centimetre of the radius as found by the Pressler's Borer.

$$\text{Increment per cent } (p) = \frac{400}{nD}$$

Where 'D' is the diameter at breast height.

'n' is the number of the rings in the last centimetre or inch.

- A fast growing species is one which yields a minimum of **10 m³ /ha/ year**. In case of younger plantation, the height increment must not be less than **60 cm per annum**.
- World average mean annual increment is **2.1 m³/ha**.
- Mean annual increment of ***Indian forest*** is **0.7-0.9 m³/ha.**
- Mean annual increment of a fast growing species is **10 m³/ha.**
- The increment in the value per unit volume of a tree or a crop, independent of any increase in the price of forest produce resulting from any change in money value in general, or the supply and demand position in particular called **Quality Increment**.
- The increment in price independently of quality increment, resulting from any change in the market on an account of change in money value in general, the demand and supply position in particular called **Price Increment**.
- **Price Increment** and **Quality Increment** can be determined and expressed in the same way as volume increment percent by using **Pressler's method**.
- The most accurate method consists in calculating the ***periodic mean increment*** from the difference between two measurements. This is suitable for calculating the increment not only on the whole stand but also by size-classes.

- Increment from C.A.I. can be calculated by using Area Method, Per Tree Method and by Increment Percent method.
- Determination of increment in **irregular crop** is done by using Andre's formula.
- Biolley's check method is used to determine increment and based on successive inventories.

  According to Biolley the M.A.I. = $\dfrac{(V_2 - V_1) + N - P}{n}$

  V_1 = initial volume of growing stock

  V_2 = final volume of growing stock

  N = material removed during the period (thinning etc.)

  n = period between successive enumeration.
- An **age-class** is one of the intervals into which the range of age of the trees growing in a forest is divided for classification or use.
- The unit of Yield Regulation in both regular and irregular forests is a ***felling series***.
- The total volume of trees in a fully stocked forest with normal distribution of age-classes for a given rotation is called **normal growing stock (N.G.S).**
- At the **beginning of the growing season** the volume of the growing stock will be:

  $$\text{N.G.S} = \left(I \times \frac{r}{2}\right) - \frac{I}{2}$$

  At the **middle of the growing season.**

  $$\text{N.G.S} = \left(I \times \frac{r}{2}\right)$$

  At the **end of the growing season** normal growing stock will be.

  $$\text{N.G.S} = \left(I \times \frac{r}{2}\right) + \frac{1}{2}$$

  Growing stock obtained from the yield table is more accurate than that obtained from the M.A.I formula.
- Flury suggests that in the mid-season formula a variable constant 'C' should be substituted for 1/2, to obtain more correct results called ***Flury constant***. Value of Flury constant 'C' will be less than 1/2 for short rotation and more than 1/2 for long rotation.

- **Fischer** evolved a formula for a condition when a part of the growing stock has been removed under regeneration felling. The N.G.S under such working would be:

$$\text{N.G.S} = (V + V_1) \times \frac{P}{2} \times D$$

  V = initial volume of growing stock

  V_1 = growing stock at the end of regeneration period

  P = regeneration period & D = crown density

- For the determination of N.G.S in selection system **Munger** evolved a formula which was based on C.A.I.
- The per cent ratio of normal yield to the normal growing stock is called **utilization per cent**.
- **Yield table** is constructed assuming sample plots are fully stocked i.e., **unity (1.00)**.
- Ratio to M.A.I of the quality to be reduced to M.A.I of the standard quality is known as ***reducing factor for quality***.
- The diameter or girth decided upon as the normal size for felling in order to fulfil the objects of management is called **exploitable size**.
- Bulk of our Indian forest (**about 70%**) are irregular in diameter and density distributed.
- According to ***De Liocourt's law*** in a fully stocked selection forest, the number of stems falls off from one diameter class to the next in geometrical progression, which means that the percentage reduction in the stem number from one diameter class to the next is constant.
- Meyer simplified ***De Liocourt's law*** in the form of an exponential function, which is given below:

$$Y = Ke^{-ax}$$

  Y = number of stems in the diameter interval, height

  x = diameter at breast, a = percentage reduction in the number of stems for each diameter class. e = 2.71828

  K = relative stand density which dependant on site conditions.

- The volume or number of stems that can be removed annually or periodically, or the area over which fellings may pass annually or periodically, consistent with the attainment of objects of management is called **yield**.

- All the material that counts against the prescribed yield and which is derived from the main fellings in a regular forest is called **final yield**.
- All material from thinning or operations preceding the main felling in a regular forest, or its cash equivalent is called **intermediate yield**.
- The yield from a normal forest is called **normal yield**.
- The material that a forest can yield annually in perpetuity is called **sustained yield**.
- The standing volume of a crop plus the total volume removed in the thinning since its establishment, as a more or less even-aged stand; or the sum of the final and intermediate yields is called **total yield**.
- The total quantity of material per annum, of given species, that an area is capable of producing under normal conditions, so long as the factors of locality remain unchanged is called **yield capacity**.
- The calculation of the amount of material which may be removed from a forest, annually or periodically over a stated period, or of the annual or periodic area over which fellings may be made, consistent with the treatment prescribed is called **yield determination**.
- A term generally applied to the determination of the yield and the prescribed means of releasing it, called **yield regulation**.
- Objects of **yield regulation** are given below-
    1. To cut each crop or tree at maturity.
    2. To obtained maximum yield of the desired produce.
    3. To cut, approximately, the same quantity of material annually or periodically.
    4. To limit the area to be felled, that can be regenerated.
- The yield is usually regulated for the period of working plan i.e., **ten years**.
- Thinning, pruning, cleaning and improvement fellings are the examples of **intermediate yield**.
- **European Continent** is the birth-place of scientific forestry.
- **S/α/M/0.6** shows middle aged (M), Sal (S) forest, quality, with a density **0.6**.
- Two stage stratified random sampling method is used for **bamboo** enumeration.
- Period of Working Plan is **10** years in India.

- The forest survey of India (FSI) conducts forest cover survey once in every **two years.**
- **Working Plan Officer** is a special duty officer at the rank of Deputy Conservator of forest. Working Plan Officer should draw up the working plan.

Basis	Method
A. Area	A1- Annual coupe by gross area A2- Annual coupe by reduced area
B. Volume	B1- Von Mental's Formula B2- Howard's modification B3- Simons' modification B4- Smythies modification B5- Burma modification
C. Area & Volume	C1- Permanent Periodic block method C2- Revocable Periodic block method C3- Single Periodic block method C4- Floating Periodic block method C5- Judeich's Stand Selection Method
D. Increment	D1- Increment method D2- Swiss method D3- Biolley's Check Method
E. Volume & Increment	E1- **Formula methods** 1. Austrian 2. Heyer's 3. Karl's 4. Hundeshagen's 5. Breymann's E2- Hufnagal's methods E3- 1. French Method 2. Melard's method 3. Smythies modification 4. Chaturvedi's modification E4- Hufnagal's **Diameter class** method E5- Brandis' **Diameter class** method or The Indian method E6- Volume unit method E7- Smythies' Safe-guarding formula or U.P Safe-guarding formula

- **B2, B3, B4, E5 and E7** are purely of Indian origin.
- Annual coupe by gross area (A1) is the oldest and simplest form of yield regulation in regular forest. It was first applied in **France** in the **14th century**.

- In annual coupe by gross area (A1) method the area of forest or the felling series is divided into a number of annual coupes to the number of years in a rotation.
- In India this method was widely adopted for **fuel wood plantation**.
- **Cotta's formula** for annual yield determination is:

$$Y_a = \left(\frac{V}{P}\right) + \frac{I}{2}$$

  Y_a = annual yield, P = period, V = standing volume in P.B.I, I = annual increment of PBI.
- In **revocable periodic block method** rotation is divided into suitable regeneration periods and the fellings series into corresponding periodic blocks.
- Floating Periodic block method is also called as **Quartier Bleu Method.**
- Formula of **Duchafour's Crown Cover** which is used to calculate the modified or reduced area (C) is given below:

$$C = K^2 g^2 n$$

  Where, g = mean girth/diameter of the trees in group, K = coefficient, and n = number of the trees in group.
- **Floating Periodic block** method is not suitable for irregular forests.
- Forest is divided into a number of compartments of suitable size in **Judeich's Stand Selection Method**. The essence of this method is careful selection of mature stand.
- **Von Mental's Formula** (based on volume) for the determination of annual yield is given as:

$$Y_a = \frac{2\ \text{N.G.S}}{r}$$

  Where, r = rotation period. Y_a = annual yield.
- Von Mental's Formula is also known as formula of **glorious simplicity**. This formula is best applicable to regular, even-aged or nearly even-aged normal forests.
- **Mason's formula** is identical with that of Von Mental's Formula. Mason's formula also known as **exploitation percent or Masson's ratio.**
- **Howard's modification** was evolved to meet the special case of selection forest in which enumerations were done down to half the exploitable girth. Howard's formula for yield is:

$$Y = \frac{8V}{r}$$

V = volume of growing stock enumerated down to half rotation age.

- **Simons' modification** is applicable to all fractional enumeration of G.S, down to girth/diameter, equivalent to 1/n th of the rotation while in case of Howard's modification enumerated is carried out down to the diameter corresponding to half the rotation.
- The **formula** of Simons' modification is:

$$Y_a = \frac{2n^2}{r(n^2 - 1)} \times V$$

- **Smythies' modification** is the modification of Von Mental's Formula. The formula of Smythies modification is given below:

$$Y_a = \frac{2V}{r - x}$$

Where, x = age down to which enumerations is done.

- Like Howard's modification, **Burma modification** was also evolved as a special case in which enumerations and measurement of the G.S was done down to one third rotation, instead half the rotation in Howard's.
- The **formula** of Burma modification is given below:

$$Y_a = \frac{9V}{4r}$$

- Historically, **Austrian method** was the first method which was based on the yield on a knowledge of increment and G.S. The formula of Austrian method is given below:

$$Y = i + \frac{(V_a - V_n)}{r}$$

Where, V_a = the actual G.S, V_n = the normal G.S, r = the period during which the difference between the normal and actual G.S is to be adjusted.

- Austrian method is applicable to both regular and irregular forests.
- **Heyer's modification** was based on actual mean annual increment. This formula was tried in Thanu forests of Dehra Dun by **M.P. Bhola**. The **formula** of Heyer's is given below:

$$Y = I_a + \frac{(V_a - V_n)}{x}$$

Where, I_a = actual increment of forest. X = the number of years in which the surplus or deficit is to be adjusted.

- **Hundeshagan's method** was based on that actual yield (increment) bears the same proportion to the actual (real) growing stock (G_r) as the normal yield does to the normal growing stock (G_n). The **formula** of Hundeshagen's is given below:

$$Y_r = G_r \times \frac{Y_n}{G_n}$$

- The quotient is called **Utilization factor by Hundeshagan**.
- **Breymann's method** is based on the same principle as Hundeshagan's method. The formula of Breymann's is given as:

$$Y = Y_n \times \frac{A}{A_n}$$

Where, Y_n = normal yield, A = average age of the fellings series, A_n = normal age of the fellings series.

- **French method (1883)** is sometimes called as **Quartier bleu** method of yield regulation in irregular forests. It was primarily designed for regulating the yield of tolerant species. The formula of is given below:

$$Y_a = \frac{V_0}{r/3} + \frac{V_0 \quad t_1}{2}$$

Where Y_a = annual yield, V_o = volume of old class, t_i = increment per unit volume of old class, r = rotation.

- In French method rotation '*r*' is divided into three equal parts viz. young, medium and old in the ratio **1 : 3 : 5**.
- In French method the volumes of old and medium classes is in the ratio of **5 : 3**. This method is simple when the proportion of old and the medium wood is **5 : 3** or nearly.
- If $V_o > V_m$, then the crop in V_o is examined and transfer **lower diameter** trees to V_m the proportion of 5 : 3 is adjusted.
- If $V_o < V_m$, then the crop in V_m is examined and transfer **upper diameter** trees to V_o the proportion of 5 : 3 is adjusted.
- If $V_o + V_m$ is less than normal (5 : 3), in this case practically nothing should be removed from V_o except the dead and dying trees, and nothing will be removed from V_m.
- If $V_o + V_m$ is greater than normal (5 : 3), the only possibilities in this case is, if $V_o > V_m$ then excess in the G.S over the normal has to be removed.

- 1894 modification of French method is called **Melard's formula.** This 1894 modification of French method was suggested by Melard. and so is called Melard's formula.

$$Y = \frac{V_0}{r/3} + \frac{V_0 \times t_1}{2} + \frac{1}{n} \times (V_m \times t_2)$$

Where, t_2 = increment per unit volume per annum in V_m, n = varying factor.

- Sometimes Melard's method is also known as **regeneration area method**.
- **Chaturvedi's formula** for annual yields is:

$$Y = \frac{V_0}{R - \mathrm{r}} + \frac{1}{2}(V_0 \times i)$$

Where, i = increment per unit volume per annum of the V_m.

- Increment method is based on the premises that yield must be correlated with the annual increment throughout the life of the crop. The formula is:

$$Y = \frac{V + a - V_n}{r}$$

Where, V = volume percent, V_n = volume 'n' years ago, a = volume removed during the previous 'n' years.

- Increment method is simplified form of **Biolley's Check Method**.
- According **Swiss method** only the annual increment is to be removed from the oldest diameter classes as the yield.

$$\text{Annual yield} = \frac{V + Z}{c.c}$$

$c.c$ =cutting cycle, Z = surplus G.S, Z = above diameter class volume.

- The principle behind the method is that the composition and distribution of the G.S can be moulded; and., by correct harvesting the timber production can be improved. This method called is **Biolley's** x**"Check method" or Method-de-Controle**. Formula of his method is give below:

$$\text{Annual increment or yield} = \frac{(V_2 + N) - (V_1 + P)}{n}$$

V_1 = G.S volume 'n' years ago, N = volume of tree cut, P = volume of recruitment in 'year.

- Biolley's "Check method" or Method-de-Controle with suitable modification, may be adopted for valuable selection forest of **Sal** and **Teak** in India.
- In 1893, Hufnagl devised another method on the basis of diameter at which tree are marketable, instated of age, for use in selection forests. This is known as **Hufnagal's Diameter class method**. The formula that he used for calculating the annual yield was as follow:

$$Y_a = \frac{n_4}{a_4 \cdot a_3} \times V_4 + \frac{n_3 - n_4}{a_4 \cdot a_3} \times V_3 + \frac{n_2 - n_3}{a_3 \cdot a_2} \times V_2 + \frac{n_1 - n_2}{a_2 \cdot a_1} \times V_1$$

  a_1, a_2, a_3 and a_4 are the mean age of trees in each dia. Classes.

  n_1, n_2, n_3 and n_4 are number of trees in lowest to the highest dia. class (in ascending order)

- Brandis' **Diameter class** method or The Indian method is based on Tree as a unit. Yield in Brandis' method is based on the number trees in various diameter/girth classes.
- Brandis' **Diameter class** method or The Indian method is suitable for extensive tropical forests where only one or two species are sealable and where *selection fellings* are the rule.
- Smythie's Safe-guarding formula is also known as U.P Safe-guarding formula. It was formed mainly for sal forest in U.P. formula is given as:

$$X = \frac{f}{t}(\text{II} - Z\%\text{II})$$

  f = fellings cycle, t = time taken by a class tree to pass into class

  Z = % of class trees that do not pass into class in 't' years.

  X = number of selected trees per ha.

- **Smythie's Safe-guarding formula** is now widely used in *selection forest* of India. Fir and Spruce forests of Chakrata Forest Division are being under selection system and yield is regulated by Smythies' Safe-guarding formula.
- Some important forest maps with their features are shown in tables.

Name of map	Scale of map	Map showing
Management map	1 : 50000	New Working Circle, felling series, periodic blocks etc.
Stock map	1 : 15000	Distribution of forest type, main forest species, non-forest area, blanks etc.

Regeneration survey map	1 : 5000	Compartment History Files and cultural operations etc.
Working Plan map	1 : 50000	Physiographic features, block name & boundaries, compartment, Forest Rest House, roads, features of reserved forest,
Enumeration map	1 : 50000	Plot location, strips, topographic units, sub-compartment etc.
Forest type map	1 : 50000	Forest types
Reference map	Size & shape of Division	Main boundaries of forest, ranges, roads, canals, F.R,Hs, neighboring towns and village etc.
Soil map	Area	Broad soil types

- Table showing forest group with their specific colour. According to Champion and Seth forest classification, there are five major forest types and 16 are major forest groups.

Group Designation	Colour
Tropical Wet Evergreen	Violet
Tropical Semi-Evergreen	Purple
Tropical Moist-Deciduous	Cobalt blue
Littoral & Swamp forest	Prussian blue
Tropical Dry Deciduous	Yellow
Tropical Thorn	Red
Tropical Dry Evergreen	Pink
Sub-tropical Broad–leaved Hill forest	Vermillion
Sub-tropical Pine	Orange
Sub-tropical Dry Evergreen	Burnt umber
Northern Wet Temperate	Olive green
Himalayan Moist Temperate	Green
Himalayan Dry Temperate	Yellow
Sub Alpine	Burnt sienna
Alpine Scrub forest	Sepia
Alpine	Sepia

Cumulative weightage value	Regeneration status	Colour of regeneration map
0-20	Deficient	Blank
21-40	Fair	Red
41-60	Moderate	Yellow
61-80	Good	Blue
81-100	Excellent	Green

CHAPTER 4

Wildlife Management

- **Wildlife comprises** of all living organisms such as plants, animals, micro-organisms in their natural habitats which are neither cultivated or domesticated nor tamed.
- **Wildlife management** is the science and art of changing the characteristics, an interactions of habitats, wild animal population, and mean in order to achieve specific human goals by means of wildlife resources.
- Book **"Wildlife Management Techniques"** was written by Robert H. Giles in 1984.
- Rewa in Madhya Pradesh is famous for the **White tigers.**
- Largest animal in the world of family Bovidae is **Gaur (*Bos gaurus*)**.
- **Renewable natural resources** are those which are regenerated through natural cycle. Such as air, water, soil, plants, animals and wind etc.
- **Non-renewable natural resources** are those which are not replaced in the environment after their utilization. Such as metals, coal, natural gas and minerals etc.

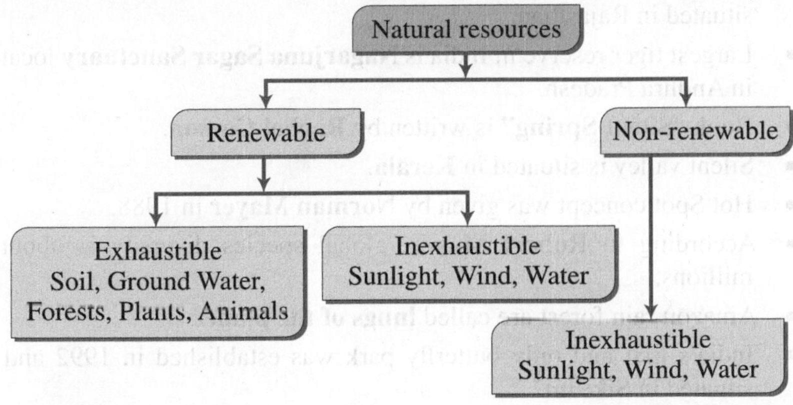

- First biosphere reserve of India **Nilgiri Biosphere Reserve** was declared in 1986.
- Biosphere Reserve has following four zones-
  1. **Core zone:** It lies at the centre where no human activity is allowed.
  2. **Buffer zone:** Where limited human activity is allowed.
  3. **Manipulated zone:** Human activity is allowed but ecology is not permitted to be disturbed.
  4. **Restoration zone:** Degraded area for restoration to near natural form.
- First wildlife conservation movement in India was started by **Bombay Natural History Society (B.N.H.S.)**.
- Predators have an important role in regulating "**balancing of nature**."
- There are four types of **predation** viz. chance predation, habit predation, sanitary predation and sucker-list predation.
- **Population-dynamics** is the study of analysis of the changes in the numbers of animals in wildlife population at a particular time.
- Conservation of fauna and flora at their native place is called ***In-situ* conservation.**
- Conservation of fauna and flora outside their native place is called ***Ex-situ* conservation**.
- World's first national park is **Yellow Stone National Park** situated in America.
- India's first national park is **Jim Corbett National Park** situated in Nainital, Uttarakhand.
- Smallest tiger reserve in India is **Ranthambore National Park** situated in Rajasthan.
- Largest tiger reserve in India is **Nagarjuna Sagar Sanctuary** located in Andhra Pradesh.
- Book "**Silent Spring**" is written by **Rachel Carson**.
- Silent valley is situated in **Kerala**.
- Hot Spot concept was given by **Norman Mayer** in 1988.
- According to **Robert Mayer**, global species diversity is about 7 millions.
- Amazon rain forest are called **lungs of the planet**.
- India's first and only butterfly park was established in 1992 and is situated in **Sikkim**.

- The modification done in the habitat as per the requirement and benefit of the wildlife especially for food, water, shelter and area is called **habitat-manipulation.**
- Salt lick is a common methods of artificial-feeding to the wild animals in pinch periods. Artificial salt-licks are made by mixing the normal salt (NaCl) with licking soil.
- In Betula National Park of Jharkhand during night a powerful torch with coloured papers is used giving signals like "**Red**" means fire is on, "**Blue**" means fire is adjoining jurisdiction, "**Green**" means all clear.
- The animals whose body temperature does not fluctuate with the changes in the temperature of the environment and are able to regulate and maintain the body temperature at a constant level called **eurythermal** or **homoeothermic** or **endothermal** or **warm blooded animals**. Such as birds and mammals.
- The animals whose body temperature fluctuate or varies with the changes in the temperature of the environment called **Poikilothermic** or **stenothermal** or **ectothermal** or **cold-blooded animals**. Such as pieces, amphibians and reptiles etc.
- According to **Bergmann's** rule the animals living in cold region are much longer than the warmer region.
- According to **Jordan's** rule the fishes living in low water temperature have more vertebrae than those living in warm water.
- Animals which can tolerate narrow fluctuation of the salt-concentration is called **stenohaline.**
- Animals which can tolerate wide range fluctuation of the salt-concentration is called **euryhaline**.
- The most tasteful and liked food is called **preferred-food** of the species.
- After preferred food, food which is important for the living of the animal being for long time called **staple food**.
- When there is scarcity of staple food, the species depends upon certain food which is neither so tasteful nor nutritive called **emergency food**.
- The food which is totally non-nutritive and is consumed by the animals only to fulfil its stomach is called **stuffing food**.
- The period in which the food is not sufficiently available and causes trouble to the animal, is called **pinch period**.

- The place or area which gives protection and serves other biological needs of the species, is known as **shelter**.
- **Cover** are those vegetation or plants which save the animals from the causal-factors and provide safety, resting, shades as well as protecting from adverse environmental effects.
- Some examples of recently **extinct animals** are dodo (Mauritius), quagga (Africa), thylacine (Australia), stellars sea cow (Russia) and three subspecies (Bali, Javan, Caspian) of tigers from the world.
- Table showing names of some important wild animals:

Common name	Scientific name/Zoological name
Lion	*Panthera leo*
Indian Lion (Asiatic lion)	*Panthera leo persica*
Tiger	*Panthera tigris*
Indian Tiger	*Panthera tigris tigris*
Panther/ Leopard	*Panthera pardus*
Snow leopard	*Panthera uncia*
Caracal	*Felis caracal*
Mangoose	*Herpestes edwardsi*
Hoolock	*Hylobates hoolock*
Jackal	*Canis aureus*
Indian fox	*Vulpes vulpes*
Sloth bear	*Melursus ursinus*
Himalayan Black Bear	*Selenarctos thibetanus*
Striped Hyaena	*Hyaena hyaena*
Red Panda	*Ailurus fulgens*
Gaur	*Bos gaurus*
Lion tailed Macaque	*Macaca Silenus*
Golden Langur	*Presbytis geei*
Wild buffalo	*Bubalus bubalus*
Wild Goat	*Capra hircus*
Indian Goat	*Capra capra*
Himalayan Thar	*Hemitragus jemlahicus*
Nilgiri Thar	*Hemitragus hylocrius*
Nilgai	*Boselaphus tragocamelus*
Four horned Antelope	*Tetracerus quadricornis*
Chinkara	*Gazella gazella*

Hangul	*Cervus elaphus hanglu*
Sambhar	*Cervus unicolor*
Swamp deer	*Cervus duvauceli*
Musk deer	*Moschus chrysogaster*
Indian wild Boar	*Sus scrofa*
Blue whale	*Balaenoptera musculus*
Gangetic Dolphin	*Platanista gangetica*
Asiatic wild Ass	*Equus hemionus*
Indian Elephant	*Elephus maximus*
One Horned Rhinoceros	*Rhinoceros unicornis*
Indian Pangolin	*Manis crassicaudata*
Blackbuck	*Antelope cervicapra*
Cheetal	*Axis axis*
Indian Porcupine	*Hystrix indica*
Common langur/hanuman	*Presbytis entellus*
Wolf	*Canis lupus*
Ibex	*Capra ibex*

- **Horns** are formed of two parts viz. an internal bony structure and outer hollow cover.
- **Antlers** are entirely solid structure, can regrow, branched and found only in male.
- Table showing names of largest/smallest wild animals:

Mammals	
Largest and heaviest animal	**Blue whale**
Fastest animal	Cheetah
Largest land animal	African Bush elephant
Slowest animal	Three toed sloth
Smallest land animals	Kitti Hog-nosed Bat
Largest Deer	Sambhar
Smallest Deer	Mouse deer
National animal	Tiger

Birds (Aves)	
Largest bird	Ostrich
Smallest bird	Bee-humming
National bird	Peacock
Reptiles	
Largest and heaviest	Estuarine Crocodile
Largest lizard	Komodo Dragon
Tree snake	Dryophis
Amphibians	
Largest amphibians	Chinese Giant Salamander
Largest frog	Goliath frog
Pisces (Fishes)	
Largest	Whale Shark

- Table showing International/National organizations concerning wild conservation:

Organization	Year	Headquarter
CITES	1976	Washington
GTF (Global Tiger Forum)	1993	New Delhi
WWF (Worldwide Fund for Nature -1990)	1961	Gland, Switzerland
BNHS (Bombay Naturel History Society)	1983	Bombay
WPSI (Wildlife Preservation Society of India)	1958	Dehra Dun, India
IBWL (Indian Board For Wildlife)	1952	-
IUCN (Now as WCU-World Conservation Union)	1948	Gland, Switzerland
CBSG (Conservation Breeding Specialist Groups)	1994	-

- Table showing location of national parks and found important wild animals:

Name And Location	Important Animals Found
Kaziranga National Park (Assam)	Rhinoceros
Sunderbans tiger reserve (W.B)	Tiger
Dachigam National Park (Kashmir)	Hangul
Hazaribagh National Park (Jharkhand)	Tiger

Corbett National Park (Uttarakhand)	Tiger
Kanha National Park (M. Pradesh)	Tiger, panther,chital,Chinkara, four horned deer
Tandoba National Park (Maharashtra)	Tiger
Bandipur National Park (Karnataka)	Elephant, tiger and leopard
Desert National Park (Rajasthan)	Great Indian Bustard and Blackbuck, chinkara
Gir National Park (Gujarat)	Asiatic lion.
Dudhwa National Park(Uttar Pradesh)	Tiger

- The number of animals per unit area is called **density**.
- The phenomena of regular movement of a species from one place to other and back is called **migration**. It is most common in birds. Migration of Siberian crane from Russia to Bharatpur (Rajasthan, India) is world famous.
- The phenomena of movement of a species from one place to other permanently is called **dispersal**.
- The status and position of distribution of the individuals of a species in the particular habitat is called **dispersion**. It is three types viz. random, uniform and clumped dispersion.
- When the individuals of a species move into an area leaving its original habitat is called **immigration** (movement into an area).
- When the individuals of a population move out from its original habitat permanently is called **emigration** (movement out of an area).
- The rate of birth in a population is called **natality**. The rate of death of individuals in a population is called **mortality**.
- **Fertility** refers to the bearing of young ones that is "occurrences of births".
- **Fecundity** is used to indicate the capacity to bear young ones.
- Animals living singly is called as **loner or solitary**. It is found in lion.
- The species which are in danger of extinction and whose survival is unlikely if the causal factors continue to be operating are called **endangered species**.
- The species likely to move into the endangered categories in the near future if the causal factors continue to be operating are called **vulnerable species**.

- The species with small population in the world at restricted area is called **rare species.**
- A book containing a record of animals & plants which are known to be in danger is called **red data book** (1948). This book is maintained by IUCN.
- A book containing a list of rare plants in a protected area like as Botanical garden is called **green data book.**
- The Burning Forest book was written by Nandini Sunder.

CHAPTER 5

Forest Utilization

- **Axes, saws** and **wedges** are the conventional implements used in felling and conversion.
- The axes are used for **felling, trimming, splitting and grubbing**.
- Axes consist of two parts- **iron head** and **wooden handle.**
- Axe head carries a socket called the **eye**, into which the handle fits.
- The portion of the iron head infront of the eye is known as the **blade**.
- For hardwood blade of an axe should be comparatively lighter and thinner than for softwoods.
- The weight of an ordinary felling axe head varies from **0.7 kg to 1.8 kg**.
- Generally the weight of the axe is 1.5 to 2.0 kg.
- In India **round handles** fitting in round eyes are almost always used.
- European and American axes usually have **oval-shaped handles** and eyes.
- **Bill-hook** is used for felling bamboo and small poles and for cutting brushwood.
- The **saw** consists of a blade toothed at one edge and one or two handles attached to the ends of the blade.
- Saws are used for **felling, crosscutting and ripping**.
- The great advantage of saws over axes in felling and conversion is saw cause much less **wastage of wood**.
- There are three popular kinds of saws used for felling and conservation-
  a. M-tooth saw
  b. The triangular or Peg-tooth saw
  c. Raker-tooth saw
- Saws are of two main types:
  a. Crosscut saw
  b. Ripping saw

- There are three main types of ripping saws used for conservation of scantling and planks.
  1. The frame saw- for Himalaya forests.
  2. The Delhi saw- for cutting sal in Uttar Pradesh.
  3. The pit saw- convert large logs into beams & used in South India.
- **Face**- Edge of the tooth which faces the cutting direction.
- **Space**- Distance between two adjacent teeth.
- **Gullet**- Entire opening between two adjacent teeth.
- **Pitch**- Angle tween the face of a toot and the line passing through points of the teeth.
- **Gauge**- The thickness of the saw blade.
- **Kerf-** Width of the saw cut.
- **Back**- The opposite end of the teeth.
- Crosscut double handed saw vary from about 140 to 250 cm in length.
- Crosscut saw also have triangular shape teeth called rakers.
- Saws should always be greased to avoid rust.
- There are two methods of saw teeth setting, **spring-setting** and **swage-setting**.
- Peg-toothed saw consists of all teeth in the form of isosceles triangles. Peg-toothed saw is the most common toothing.
- The **'cutters'** cut the wood fibres along both sides of the groove during cutting the log. While **'Raker tooth'** works like a chisel and break off the fibres.
- The techniques of proper levelling of the saw teeth is called **jointing**.
- **Wedge** are used for splitting logs or fuel billets, and for assisting in the felling of trees and longitudinal sawing of timber.
- The metal wedges are made of steel or iron and are generally **30 cm** long, **8 cm** wide and weigh upto **5 kg**.
- The height of the stumps should be as low as possible and not more than **15-30 cm**.
- The felling axe is slightly heavier than trimming axe.
- The standard length of axe handle is **90 cm**. The length of axe handle varies from **70-120 cm**.
- The length of crosscut saw varies from **160-220 cm** and width from **8.6 to 12.2 cm** in the centre.

- **Bow saw** is very useful for a single man cutting up to diameter 20 cm.
- **Cant-hooks & pickaroons** are used as lever in rolling, stopping and turning logs. The length of cant-hook handle varies from 1.5 to 1.8 m.
- **Barking** spade is used for removing the bark from a felled tree while it is in green condition.
- **Cleaver** is use for splitting firewood & also driving is wedges.
- A meter-long wooden stick made of seasoned timber is called **measuring stick**.
- **Cable-tensioner or cable-puller** is an apparatus developed in Switzerland, is used to fell the tree in desired direction to minimise the damage of another tree. With this apparatus, two or three men can exercise a pull of **3.5 tonnes**.
- **Stem-tightener** consist **13 mm** wire-rope cable which is put around the stem, just above the felling cut after removing the bark. It is used to prevent the butt ends of trees, especially leaning ones, from splitting during felling.
- **Power chain saws** are not much used in India. These saws are operated by petrol or electricity. Power chain saws are much used in western countries.
- The lengths of the standard blades of power saw is **50-80 cm**.
- The best season for felling is **winter season**. In the plain and submontane tracts of India, the felling season is from October to March.
- There are two main principles underlying all methods of felling:
  1. The production of the maximum amount of sound timber that can be made available for export form the forest.
  2. The avoidance of damage to the surrounding forest.
- In order to satisfy above conditions the following rules must be observed:
  a. Tree should be felled as near the ground as possible.
  b. Trees should be felled in a manner and in the direction in which they will do least damage to themselves and the surrounding forest.

     On **hilly ground** it is the best to fell trees **uphill side**.
  c. Tree should not be felled such that they fall into a place where it is not possible to convert or extract the timber.

d. Trees should be felled in a manner and in a direction in which they will do least damage to surrounding crops.

e. Trees should not be felled during a strong wind.

f. Felling should usually begin at the top of a slope and proceed in a downward direction.

g. Felling should be concentrated as much as possible.

- There are three important methods of felling a tree with axes and saws.
  - With the axe alone
  - With the saw alone
  - With the saw and axe combined
- Felling with saw and axe is mostly confined to **Jammu and Kashmir, Punjab, Himachal Pradesh and Uttar Pradesh**.
- During felling with axe alone an **undercut** is given with the axe towards the tree is to fall, upto two-third of the tree diameter. While the **felling cut** is then made exactly opposite to undercut about 10 to 15 cm above the undercut.
- During felling with axe and saw combined, a cut extending to about one-fifth or one-fourth of diameter of stem is first made with axe on the side to which tree is intended to fall. Then a saw cut is made on the opposite side of axe cut. This method is most satisfactory as direction of fall is easily controlled.
- During felling with saw alone, **cut** is given opposite side to which tree is intended to fall. This method involves the least wastage of wood. This method is seldom used in India.
- Felling by root is always desirable in the case of very valuable trees such as **sandal tree** or **khair tree** and **walnut tree** etc.

Name	Gauge (in mm)	Standard Sleeper Size (in mm)
Broad Gauge	167.64	275 × 26 × 13
Metre Gauge	100.00	183 × 21 × 19
Narrow Gauge	76.20	153 × 18 × 12
Light Narrow Gauge	61.00	153 × 16 × 11

- Table showing wood wastage during various logging operations:

S.No.	Wood wastage form	Wastage percentage
1.	**Felling:** Stumps - Damaged to surrounding tree by felled tree - Stem splitting - Felling of leaning trees without any control	4-6 2-3 1-2 2-3
	Total wastage due to all fellings	9-14%
2.	**Sawing:** Hand sawing Pulp wood	27-32 4-5
	Total wastage due to sawing	31-37%
3.	**Minor transportation** - Chutes and dry slides - Wet Slides - Telescopic floating	1 2-3 2-3
	Total wastage due to minor transportation	5-6%
4.	**Major transportation** - Normal river loss	3-5
	Total wastage due to transportation	8-11%
	Grand total of wastage	48-63%
	Total average wastage	56 %

- The cost of felling, conversion and transportation is approximately **50%** of total cost of delivery of timber in the market.
- The total volume of usable rough timber from the average tree is only **37-52%**.
- First of all in India logging work was started by Dr. **A. Huber** and **A. Koroleff**.
- First Logging branch in India was established in **June, 1957** in the **Forest Research Institute (F.R.I)-Dehra Dun** during 2nd five year Plan.
- First Logging Training Centre in India was established in 1958-59 at **Batote** (J&K).
- Nomenclature of sawn materials according their common sizes and market name is given below:
  - **Squares:** Completely squared, rectangular shape.
  - **Hakries:** The rounded log split with axe.
  - **Poles:** The rounded products of varying lengths.

- **Beams:** More than 4 meter length having end cross section more than 15 cm.
- **Scantlings:** Square or rectangular shape having 15 cm under section.
- **Planks:** Wooden slabs having maximum thickness 5 cm.

- **Hand sawing** is most common conversion method used in Indian forest.
- Minor transportation also called as **Off road transportation**.
- Transportation capacity of a motor truck is generally **1.5 to 5 tons**.
- **Floating, rafting & booms** and wet slides are the methods of water transportation.
- **Depot** is a place where wood or other forest produce is stored, pending its disposal.
- Depots can be classified into three broad categories viz., **forest depots, transit depots and sale depots.**
- Transit depots are intermediate depots inside or outside of forest. **Sale depots** are located near the market.
- Sales of forest produces may be grouped into two heads viz., first is **lum-sum sale** and other is **payment on outturn**.
- Security deposit for auction is **10%** of total contract value or money.
- Forest labour is classified into three categories viz., **local labour, imported labour** and **forest villages or settlements**.
- **Suitability** of wood for various purpose depends on its technical properties such as anatomical structure, weight, strength, hardness, flexibility, elasticity, toughness, durability, colour, grain and figure.
- Table showing uses of wood:

Power	Food	Clothing	Shipping & transport	Packing	Engineer-ing	Education
Firewood Wood gas Wood alcohol	Hydro-lysed wood Wood sugar Food yeart Liquor	Rayon Staple Fibre	House Railway Carriage Truck bodies Ships Aeroplane Furniture	Packing cases Crates Teachest Boxes Ammuni-tion box	Agricultural Mechanical Chemical Electrical Machine parts	Paper Books Photograph paper Board Pencils

- Table showing types of wood most suited for special purpose:

Purpose	Wood qualities	Suitable species
Aircrafts timber	Sound. Knot free. Straight grain, light weight. High strength/weight ratio.	*Picea sitchensis* (sitka spruce) *Abies pindrew* (silver fir) *Picea smithiana* (spruce)
Agricultural implements	Strongest, Toughest and hardest	*Acacia nilotica (*babul*), Acacia catechu* (khair), *Anogeissus latifolia (*axle-wood*), Shorea robusta* (sal), *Ougenia oojeinensis* (sandan), *Mesua ferrea* (mesua), *Schleichera oleosa* (Kusum), *Xylia xylocarpa* (irul) etc.
Battery separators	Good matching property, permeable, straight grained, strong, knot free, free from volatile acids, tannins & resinous matters, resistance to the action of electrolyte & mineral acids.	Mostly conifers. ***Abies pindrew* (silver fir)** ***Picea smithiana* (spruce)** *Cedrus deodara* (deodar), *Cupressus torulosa (*cypress) *Pinus wallichiana* (kail) etc.
Bearings	Extreme hardness, very great strength	**Lignum vitae** obtained from *Guaiacum officinalis & G. sanctum* *Manilkara littoralis* (Andman bullet wood), *Hardwickia binata* (anjan), *Soymida fabrifuga* (rohini) etc.
Bent-wood articles (sports goods)	Straight grained, strong, knot free and well-seasoned	*Morus alba* (mulberry) *Fraxinus species* *Celtis australis* (celtis) *Salix alba*
Boot-lasts & Shoe-heels	Medium weight, moderately hard but tough, fine textured, smooth finishing etc.	*Fagus species &* *Acer species* (maple) – shoe-lasts *Dalbergia sisso* (shisham) *Adina cordifolia* (Haldu) *Abies pindrew* (silver fir) *Picea smithiana* (spruce) *Acer species* (maple) for Shoe-heels

Brushes & Brooms	Medium hardness, fine to medium fine texture, smooth finishing, easy to season, light colour etc. For ornamental brushes a good figure & colour are also desirables.	**Toilet & hair brushes-** *Chloroxylon swietenia* (satin wood) *Dalbergia latifolia* (rosewood) *Dalbergia sisso* (shisham) *Diospyros ebenum* (ebony) **Shaving brushes-** *Mitragyna parviflora* (kaim/Kadamb) **Horse brushes-** *Lannea coromandelica* (jhingan) **Paint & varnish brushes-** *Acer species* (maple) *Adina cordifolia* (Haldu) *Gmelina arborea* (ghamari) *Grewia tiliaefolia* (dhaman) **Brooms & general utility brushes-** *Cedrus deodara* (deodar) *Holoptelea integrifolia* (kanju) *Mitragyna parviflora* (kaim/Kadamb) *Mangifera indica* (mango)
Boat & ship	Strong, elastic, more durable, knot free, non-corrosive, low expansion & contraction coefficient and free from tannic acid.	**Ship-** *Tectona grandis* (teak) *Admirality teak* (Burma teak) **Boat-** *Quercus* spp. (European Oakwood) *Ailanthus excelsa* (mahogany) *Acer species* (maple)
Furniture, cabinet-making & Panelling	Good colour, strong, elastic & tough, strain resistance, good grain & figure, non-liability to crack, non-split, warp or move excessively and easy to work & finish.	*Tectona grandis* (teak) *Dalbergia latifolia* (rosewood) *Dalbergia sisso* (shisham) *Pterocarpus marsupium* (Bijasal) *Toona ciliata* (toon) *Albizia spp.*(Siris) *Betula alnoides* (birch) *Gmelina arborea* (ghamari) *Ougenia oojeinensis* (sandan) *Aesculus indica* (horse chestnut)
Matchwood industry (started-1922)	Straight grain, good fissility, strength & toughness, good white colour, free from knots, easily peelable, light weight, capacity to absorb paraffin and burn smokeless, efficient to burn.	***Populus tremula*** (European aspen) & ***Tilia japonica*** best matchwood. *Sideroxylon longipetiolatum* (lambapatti), *Bombax ceiba* (semal), *Sterculia villosa* (udal), *Trewia nudifera* (Gutel), *Mangifera indica* (mango), *Alstonia scholaris* (chatian), *Anthocephalus cadamba* (kadam), *Endospermum malaccense* (bakota).

Mathematical instruments	Fine, even textured, non-liable to crack & wrap.	*Pinus wallichiana* (kail), *Boxus sempervirens* (boxwood), *Juglans regia* (walnut), *Tectona grandis* (teak) *Toona ciliata* (toon), *Abies pindrew* (fir) and *Adina cordifolia* (Haldu)
Musical instruments	Fine, even textured, non-liable to crack & wrap, free from knots, having uniform annual rings, regular structure,	Conifer wood- ***Picea smithiana*** **Violins-** Bridge- *Acer species* (maple) Keys- *Tectona grandis* (teak) *Diospyros* spp. (ebony) Bows-*Dalbergia latifolia* (rosewood) *Heritiera minor* (sundri) **Sitars-** Body-*Toona ciliata* (toon) Neck-*Tectona grandis* (teak) Keys- *Cedrus deodara* (deodar) *Dalbergia sisso* (shisham) **Vina & Tambora-** *Artocarpus hetrophyllus* (kathal) *Pterocarpus marsupium* (Bijasal) *Gmelina arborea* (ghamari) **Guitar-** *Canrium euphyllum* (white dhup) **Harmonium-** *Tectona grandis* (teak) **Drums-** *Fraxinus* spp. (ash) *Dalbergia sisso* (shisham) *Albizia* spp. (Siris) **Piano-** *Juglans regia* (walnut) *Swietenia* spp. (mahogany)
Packing case industry	Light weight, reasonably strong, good nailing properties, preferably whitish, easily worked, cheap & plentiful, free from warping & shrinkage free from bed odour.	*Picea smithiana (spruce). Pinus wallichiana* (kail), *Abies pindrew* (fir), *Bombax ceiba* (semal), *Mangifera indica* (mango), *Trewia nudifera* (gutel), *Ailanthus excelsa (maharukh), Alstonia scholaris* (devil tree), *Tetrameles nudifera* (maina).
Pencil industry	Light weight, pink in colour, have good whittling & matching properties, fine even textured, straight & closed grain, free from knots.	***Juniperus virginiana* (pencil cedar)** *Juniperus procera* (wood balsa) *Juniperus macropoda* (juniper) *Alnus nepelensis* (alder)

Plywood industry	Wood with closed firm grain, round, free from knots & resin, interlocked fibres and free from excessive bulges, taper & buttresses.	*Dipterocarpus macrocarpus* (Hollong) *Terminalia myriocarpa* (hollock), *Mangifera indica* (mango), *Ailanthus* spp., *Toona ciliata* (toon). *Populus* spp., *Holoptelea integrifolia* (kanju), *Swietenia spp.* (mahogany), *Terminalia bellerica* (bahera),
Rifle parts & Gun stocks	Light moderately hard, closed grain, round. Non-split, seasoned wood, capable of standing up to the high-speed machine used in manufacturing process, warp or move in service.	*Juglans regia* (walnut) *Juglans nigra* (wild walnut) *Acer pictum* (maple) *Prunus padus* (cherry)
Toys	Light weight, fine even textured, having turning & carving qualities.	*Adina cordifolia* (Haldu), *Artocarpus hetrophllus* (kathal), *Dysoxylum malabaricum* (white cedar), *Miliusa tomentosa* (hoom), *Wrightia tinctoria* (Dhudhi), *Santalum album* (sandalwood), *Aegle marmelos* (bel).
Tents poles & tents pegs	Round or squid wood, strong. Mostly made of **bamboos.** For tent pegs-hard, durable & tough.	*Grewia tiliaefolia* (dhaman), Shorea robusta (sal), *Acacia nilotica* (babul), *Tamarinds indica* (imli), *Schleichera oleosa* (Kusum), *Ougenia oojeinensis* (sandan), *Heritiera minor* (sundri).
Tobacco pipes	High heat withstand capacity, non-split, strong wood.	*Dalbergia sisso* (shisham) *Dalbergia latifolia* (rosewood) *Tectona grandis* (teak) knots, Country pipes & hookahs- *Amoora cucullata, Boswellia serrate* *Dalbergia sisso* (shisham)

Sports good industry	Sound. Knot free. Straight grain, light weight. High **strength/weight** ratio, pliable, tough, shock resisting, easy to work, preferably light in colour and taking a good polish capacity.	Europe & America-***Fraxinus*** spp. (ash) In India most commonly used – *Morus alba* (mulberry)
	Billiard cues-light weight, Knot free. Straight grain, white colour or nearly so.	*Grewia tiliaefolia* (dhaman), *Diospyros melanoxylon* (ebony) & *Polyalthia fragrans* (debdharu).
	Bows & Arrows-very strong, tough & an elastic wood,	*Bow-Taxus baccata* (yew), *Grewia tiliaefolia* (dhaman), *Parrotia jacquemontiana* (parrotia), *Accacia catechu* (khair). Arrows- reeds & bamboos.
	Cricket bats, stumps & bails-light weight, preferably light in colour, reasonably strong, shock resistance.	***Salix*** spp. (willow) in India. The best cricket bats are made of the English willow (*Salix alba var. Caerulea*). **stumps & bails-** *Morus alba* (mulberry) & Celtis.
	Fishing rod	*Ocotea rodioei* (greenheart)
	Golf clubs	*Carya* spp. (south American hickory)
	Hockey sticks- light weight, preferably light in colour, reasonably strong, shock resistance amenable to steam bending,	*Morus alba* (mulberry) *Fraxinus* spp. (ash) *Celtis australis* (celtis)
	Skis (skiing)-toughness, flexibility, even-textured, straight grain,	*Morus alba* (mulberry) *Fraxinus* spp. (ash) *Dalbergia sissoo* (shisham)
	Tennis, Badminton & squash rackets-straight grain, long fibre wood with good shock resistance.	*Morus alba* (mulberry) *Fraxinus* spp. (ash)
Shoulder poles	Strong, tough & an elastic wood free from knots.	*Anogeissus latifolia* (axlewood) *Anogeissus acuminata* (yon) These are usually made from bamboo.
Shingles	Durable, free from knots, non-splitting, light in weight, adequate strength, taking paint & preservative capacity.	*Pinus wallichiana* (kail) *Pinus roxburghii* (chirpine) *Cedrus deodara* (deodar) *Abies pindrew* (fir) *Hopea odorata* (thingan)

Railway sleepers	Strength, resist to all pressure, good seasoned, shock resistance, good absorber of coal tar creosote.	*Shorea robusta* (sal) *Mesua ferrea* (mesua) *Cedrus deodara* (deodar) *Pterocarpus marsupium* (Bijasal)
Railway carriages	Sufficient strength, free from seasoning defects, availability & retention of shape.	*Tectona grandis* (teak) *Shorea robusta* (sal)
Walking sticks	Straightness & strength wood	Bamboo & canes are extensively used. *Celtis australis* (celtis), *Dalbergia latifolia* (rosewood), *Dalbergia sisso* (shisham) & *Santalum album*.
Umbrella handle	Extensively bamboos.	*Arundinaria* spp. *Dendrocalamus strictus* (male bamboo)
Batons	A very heavy and strong wood.	*Olea ferruginea* (olive) *Dalbergia latifolia* (rosewood), *Heritiera minor* (sundri) *Mesua ferrea* (mesua)

- **Cutch** (catechu-tannic acid) and **katha** (catechin) are obtained from the heartwood of *Acacia catechu* (Khair). Kheersal (white powder) also obtained from khair tree.
- There are three extraction method of cutch and katha which are given below:

Extraction method	Cutch yield	Katha yield
Country method	-	-
Modified method	10 to 12%	3 to 4% by weight of wood
Improved method or factory method	-	4 to 4.5% by weight of wood

- Cutch is soluble in both cold as well as hot water while katha is most soluble in cold water and this concept is used to extract the katha from hot concentrate solution of cutch.
- Katha is used in the preparation of **chewing pan** and in medicine as astringent & digestive. While cutch is used as **dying** and **preservative** agent in cottage industry.
- Paper-making in India was started in 1830. The first paper mill in India was established in **West Bengal**. The present (2015-16) per capita consumption of paper in India is **9.0 kg** while at global level it is **60.0 kg**.

- Raw material for paper making is given below:

Raw material	Species & used for
Bamboo (used for all varieties of paper)	Bambusa arundinacea, B. nutans, B. tudla, B. vulgaris, B. balcoona etc.
Sabai grass (*Eulaliopsis binata*)	Writing and printing paper
Soft-woods (conifers)	*Abies pindrew, Picea smithiana, Pinus roxburghii, Pinus excelsa Cryptomeria japonica etc. are used for paper furnishing.*
Hard-woods	Paper mulberry (*Broussonetia papyrifera*)

- The process of **paper making** followed five main steps:

Steps
1. Pulping - > Mechanical > Chemical – a) Alkaline process b) Acidic process > Semi-chemical
2. Pulp cleaning
3. Bleaching- Single stage Multi stage
4. Stock preparation- a) beating b) sizing c) loading d) colouring e) use of other additives
5. Sheet formation

- Table showing important features of all five paper making process:

Pulping process	Process used for
Pulping • Mechanical • Chemical • Semi-chemical	Isolation of cellulosic fibres from raw materials. **Mechanical** -Grinding of raw materials such as fibre, lignin etc. **Chemical-**Removal of lignin and fibrous raw materials from mechanical pulp **Semi-chemical** -For pulp softening
Pulp cleaning	Removal of unwanted pulp materials. In first stage we remove knots while in next stage centrifugal screening of shives, dirt & grit

Pulp bleaching	For colouring of pulp. In single stage bleaching calcium hypochlorite ($CaOCl_2$) is used and in multi stage bleaching chlorine gas (Cl_2) and chlorine dioxide (ClO_2) is used for bleaching
Stock preparation • Beating • Sizing • Loading • Colouring • Other additives	**Beating**-mechanical treatment given to fibres. This increased bursting strength, tensile strength, folding endurances, smoothness and rattle of paper while decreases tearing strength, bulk, porosity, opacity and dimensional stability. Thus physical characteristics of paper depends on beating. **Sizing**-sizing is done for making paper more or less impervious to ink penetration. *Rosin*, soap along with aluminium sulphate is used. **Loading**-addition of non-fibrous mineral matters called fillers. These improve the dimensional stability as well as brightness of paper. Common fillers used by industry are *china clay, talc powder* **Colouring**-for dye the paper to correct shade. **Other additives**-starch, vegetables gums, resin also used to give special characteristics to paper.
Sheet formation	After proper drying and calendring it is rolled over spools finally cut, sorted & packed

- Mechanical pulp differs from chemical pulp having all lignin of original wood and wood fibres in the form of bundles & fragments. While chemical pulp have no lignin and fibres contents. These are digested by chemicals.
- Difference between acidic and alkaline process of chemical pulping are given below:

Acidic process (Sulphite process)	Alkaline process (Sulphate process)
Cooking agent – calcium bisulphate + SO_2	Sodium hydroxide major cooking agent
Waste liquor not used for chemical recovery due to calcium deposition	Black liquor is sent to soda recovery unit for chemical recovery.
Only one principle acidic process called sulphite process is used.	Two principle alkaline process used as- 1. *Soda process*- Sodium hydroxide only 2. *Sulphate*- Sodium hydroxide + Na_2S

- **Oleo-resins** obtained from pines are soluble in organic solvents (benzene, alcohol and ether) but insoluble in water. Turpentine oil (essential oil) and rosin (non-volatile solid) are obtained from resins by distillation.

- **Chirpine** (*Pinus roxburghii*) is the only species regularly tapped in India. The total area of chirpine forests is estimated to be **0.9 million** hectares and annual resin production is about **48,000 tonnes**.
- Commonly used tapping method of resin is 'cup & lip' method. This is also called as **French method**. There are two types of tapping viz. light continuous tapping and heavy tapping.

Light continuous tapping	Heavy tapping
Tree girth is above 0.9 m	Tree girth is above 60 cm
Known as living method of tapping	Known as 'tapping to death'
Gives sustained yield annually	Sustainable yield not obtained

- New method of resin tapping called **'Rill method'** was developed in Forest research institute (F.R.I). Rill method is best as compare to cup and lip method. Rill methods gives **22%** more resin yields as compare to cup and lip method without loss of timber.
- Crude pine resin consists of two principal constituents a liquid known as **turpentine oil** and a solid known as **rosin**.
- The yield of rosin is **74%** by weight and turpentine is **17-20%** by weight while impurities and water amount is about **6-9%**.

Turpentine oil	Rosin
Colourless oily liquid having a strong odour	Colourless, yellow, red, brown or black.
Boiling point- 155°C & specific gravity-0.85 to 0.88	B.P-100° to 140°C sp. gavity-1.08
Insoluble in water	Soluble in organic solvent
Uses- paints, varnish and shoe polishes preparation. Its products are used in manufacture synthetic rubber, waxes, insecticides and germicides.	Uses- extensively used in soap making and sizing paper to take colours and inks.

- India is the major producer of lac which accounts nearly 70-80% of total world production. It is secreted by the lac insect, ***Laccifer lacca***.
- ***Butea monosperma*** (palas), Zizyhus mauritiana (ber) and ***Schleichera oleosa*** (Kusum) are the major commercial host plants for lac insect. While kusum produce best quality lac.

- The life cycle of lac insect is 3-4 days. During eggs hatching process in lac insect large number of lac larvae comes out from eggs by the phenomena known as **swarming**.
- There are two distinct strains of lac insects in India, called **Kusumi** and **Rangeeni**. Rangeeni contributes to nearly **80-90%** of the country production. The lac crops of Baisakhi & Aghani are generally collected when still **no** mature.
- **Shellac** is obtained from lac after proper washing (usually three washing). Shellac is water insoluble and soluble in organic solvent. It is used in moulding and in gramophone industries.
- Partially burnt or carbonized wood and the residue left behind during **destructive distillation** of wood known as charcoal. There are three methods of charcoal making viz. (a) burning in retorts, (b) burning in open pits and (c) burning in kilns.
- **Charcoal** is used for cooking and heating purpose. Good quality charcoal is a dry, black, and porous but a fairly hard substance. It is stored in gunny bags.
- The term **Minor Forest Products** includes all Forest Products other than **Major Forest Products** which consists of timber, small wood and fuel wood. **Minor Forest Products** includes grass, fruits, leaves barks, animal and mineral products found in forest and collected from them.
- The minor forest products of commercial importance may be divided into the following classes viz .a) fibres and flosses b) grasses, bamboos and canes c) grass oils d) oil seeds e) tans and dyes f) gum, resin and oleo-resins g) animal, mineral and miscellaneous products and h) drugs, spices, edible products and poison.
- **Fibres** are long, narrow pointed end having thick walls & small cavities and provide rigidity to the plant. Fibres are longest cell in plant body.
- The process of separation or extracting fibres from one another is called **ratting**.

Fibres obtained from stems		
Species name	**Fibre colour**	**Fibre uses for making**
Sterculia villosa (udal)	Whitish-pink	Elephant harness, drags ropes, raft tying
Grewia tiliaefolia	Yellow	Rope making

Hardwickia binata (anjan)	Red	Cordage & rope making
Bauhinia vahlii	-	Tying purpose
Acacia leucophloea (hiwar)	-	Fishing nets & cordage
Calotropis gigantean	-	Fishing nets & cordage
Trema orientalis	Light brown	Twines and coarse cloth & rope making
Ficus species	-	Rope making
Boehmeria nivea ('rhea' or 'ramie' fibre)	White	Rope making
Cannabis sativa	White	Ropes, cable, twines, mats, canvas,
Girardinia hetrophylla	-	Rope making, twines & coarse cloths
	Fibres obtained from leaves	
Caryota urens (kittul)	Brown to black	Rope making, fishing nets
Musa textilis	-	Rope making,
Pandanus species	-	Cordage, Rope making, fishing nets
Agave sislana (sisal fibre)	-	Cordage, Rope making
	Flosses yielding plants	
Ceiba patendra (kapok)	-	Spinning, weaving or **stuffing**
Bombax ceiba (Indian kapok)	-	Cotton industry
Cochlospermum religiosum	Yellow	Stuffing mattresses, pillows, cushions and life saving belts
Calotropis gigantea	-	Stuffing mattresses, pillows,

- **Coir fibres** is coarse, stiff, buoyant and elastic used for manufacture ship's rope. Coir are obtained from *Cocos nucifera* (coconut tree).
- Oils from plants are obtained by **distillation and extraction method**.

Species name	obtained	colour	Source of	Used in
Palmrosa grass *Cymbopogon martini*	Inflorescence	Pale yellow liquid	Geraniol	Cosmetics industry, mosquito repelling ointments
Lemon grass *C. flexosus*	Whole grass	Reddish yellow	Citral (41-85%)	Scents, Cosmetics industry mosquito repelling ointments

Ginger grass *C. martini*		-	-	Soap industry
Citronella grass *C. nardus*	-		-	Low cost perfume
Khus grass *Vetiver zizaniodes*	-	-	-	Perfume, soap and cosmetics
Wood oils				
Sandalwood oil *Santalum album*	Heart Wood	-	Santalol	Perfume industry
Agar oil *Aquilaria agallocha*	Heart Wood		-	Perfume industry as a perfume retainer
Deodar wood oil *Cedrus deodara*	Wood	Reddish-brown	-	Perfume, soap and cosmetics
Pine oil *Pinus sylvectris*	Heart Wood	-	-	Medicine, veterinary disinfectant sprays, soap
Leaf oils				
Eucalyptus oil *Eucalyptus globulus*	Leaves, branch	Colour Less	-	Toothpaste, perfume, vermin-repellents etc.
Citriodora oil *Eucalyptus citriodora*	Leaves	-	citronellal	Soap & perfumery
Camphor oil *Cinnamomum camphora*	Leaves and wood	-	-	Cold remedies, insecticides and perfumery
Mint oil *Mentha arvensis*	Leaves	-	Menthol	Medicines and confectionary
Wintergreen oil *Gaultheria fragrantissima*	Leaves	-	-	Perfumery and pharmaceutical
Root oils				
Costus oil *Sassuria lappa*	Roots	Pale yellow	-	Perfume blending
Valerian oil *Valerian wallichii*	Dry Rhizome		-	Flavouring and blending
Flower oils				
Keora oil *Pandanus tectorius*	Flower	-	Keora Attar	Hair oils, perfumery and cosmetics
Cassie oil *Acacia farnesiana*	Flower	-	Cassie	Perfumery
Champ oil *Michelia champa*	Flower	-	Attar	Perfumery

- **Linaloe oil** is obtained from the heartwood *Bursera delpechiana*. It is used as a fixative in perfumery and soap.
- The only commercial wood distilling plant is situated in Bhadravati, **Karnataka**.
- During carbonization of wood **25% charcoal**, **3% tar**, **4-5% acetic acid** and **3% alcohol** is obtained through destructive distillation of wood.
- **Chir tar** is obtained from *Pinus roxburghii* (chirpine) and used for tarring of ropes. While tar also obtained from deodar tree and used in ulcer, inflated skin (Sarnais).
- There are two main methods of oil extraction from seeds. These are **expression** and **extraction**. Solvent used in extraction method are petroleum, ether and carbon disulphide.

Seed oil		
Species name	**Seed yield**	**Used in**
Aleurites fordii	Tung oil	Paints and varnish
Azadirachta indica	Margo oil	Medicinal and soap making
Madhuca butyracea	Phulwara Butter	Soap, candles and chocolate manufacture
Garcinia indica	Kokam Butter	Edible fat
Hydnocarpus kurzii	Chaulmugra	Leprosy & skin disease
Madhuca indica	Mahua Butter	Cooking purpose, hair oil, soap & medicine
Pongamia pinnata	Kanju oil	Soap and leather tanning
Schleichera oleosa	Kusum oil	Cooking, lightning and soap making
Shorea robusta	Sal Butter	Cooking, lightning and soap making
Vateria indica	Piney tallow	Medicinal and lightning purpose

- Wax are usually found on the epidermis of leaves and fruits. *Sapium sesbiferum* (Chinese tallow tree) seed produces wax. *Rhus succedonea* with *Melia azadarach* fruits produces Japan wax.
- **Tannin** are the organic substances which have the property of combining with the albumen and gelatine to form an insoluble decay resist compound. Hides and skins treated with tannins called **leather**.
- **Tans** are obtained from bark, fruits and leaf. Indian tree do not produce wood tan.

Tans obtained from barks		
Species name	**Local name (Bark)**	**Tannin per cent**
Acacia nilotica	Babul	18
Cassia auriculata	Avaram	23
Cassia fistula	Konnai	3-9
Acacia mollissima	Wattle	15
Terminalia arjuna	Arjun	20-24
Ceriops roxburghiana	Mangrove tree	20-37
Tans obtained from fruits		
Terminalia chebula	Myrobalans (most imp.)	Average 32
Terminalia bellirica	Myrobalans	
Emblica officinalis	Myrobalans	
Acacia nilotica	Babul pod	12-19
Caesalpinia coriaria	Divi-Divi pods	22-40
Tans obtained from leaves		
Anogeissus latifolia	Dhonkra	Dry mature leaves- 16 Young leaves- 55 Dried green + red leaf- 30
Carissa spinarum	Karaunda	9-11

- **Dyes** are the substances which are used for imparting colour and staining purpose. These are obtained from wood, barks, flowers, fruits and leaves.

Dyes obtained from woods			
Species name	**Known as**	**Dye colour**	**Uses**
Pterocarpus santalinus	Santalinus dye	Red	Salmon pink colour to cloth
Caesalpinia sappan	Brazilin dye	Red	Dying of cotton, silk, wool
Atrocarpus hetrophyllus	Kathal dye	Yellow	Silk cloth dying
Acacia catechu	Cutch dye	Black	Tanning of lather
Dyes obtained from barks			
Acacia species	Acacla dye	Black	Dying agent
Alnus species	Alnus dye	Red	Cloth dying
Terminalia alata	Alata dye	Black	Dying as black colour

Dyes obtained from flowers & fruit			
Mallotus philippensis	Kamela dye	Orange	Dying silk
Bixa orellana	Annatto dye	Yellow	Dying of silk, wool & calico
Butea monosperma	Dhak dye	Orange	Dying cloth
Nyctanthes arbortristis	Harsringar	Orange	Liquors colouring
Wrightia tinctoria	Dudhi dye	Blue	Colouring agent
Dyes obtained from leaf			
Indigofera species	Indigo dye	Blue	Cloth dying
Lawsonia inermis	Heena dye	Red	Hair, palms & fabric dying

- The acacia gums largely known as *'gum Arabic'* are derived from several species of the *genus acacia.*

Gum yielding tree species			
Species name	**Gum known as**	**Colour**	**Uses**
Acacia catechu	Accacia gum	Dark colour	Dying and colouring
Acacia nilotica	Indian gum Arabic	–	Calico-printing, sizing material for silk and cotton
Acacia senegal	True gum arabic	–	Medicinal purpose
Pterocarpus marsupium	Gum kino	Ruby red	As astringent in medicine
Lannea coromandelica	Jhingan gum	–	Making ink ,confectionary
Moringa pterygosperma	Moringa gum	–	Medicinal use
Sterculia urens	Katira gum	–	Used in medicine for diarrhoea and dysentery
Boswellia serrata	Salai gum	Yellow	As astringent in medicine
Anogeissus latifolia	Dhaura gum	–	Paper sizing & calico print.
Bauhinia retusa	Semla gum	–	Eating & sizing cloth, paper

- The gum and resins that exude from cracks or cuts and solidify with air exposure called **dammar**. The true dammar is obtained from *Agathis loranthifolius* (conifer tree).
- Soft form of *Vateria indica* (vellapine) resin called as **'piney resin'**, **'Indian copal' or 'dhupa'**. While when it is hard called as white dammar.

- Gum which is not soluble in water but absorb water and swell up into a mucilaginous mass called as **gum-tetracanth**. True gum-tetracanth is not produced by any Indian tree. But Katira gum is slightly inferior to true gum-tetracanth.
- The resins, oleo-resins and gum resins obtained from trees are classified into (a) **pine resins,** (b) **resins** from broad-leaved species.
- **Pine resins** are obtained from *Pinus roxburghii* (chirpine), *Pinus wallichiana* (kail), *Pinus kesiya* (khasi pine) and *Pinus gerardiana* (Chilgoza pine).
- Resins from broad-leaved species are given below:

Resin yielding broad-leaved species		
Broad leaved species	**Resin known as**	**Resin uses**
Canarium strictum	Black dammar	Varnishes, bottling wax, boat chaulking
Hopea odorata	Rock dammar	Wound medicine, varnishes, paintings
Vateria indica	White dammar	Varnishes, sores ointment, boat chaulking
Shorea robusta	Ral	Fumigant, disinfectant, varnishes & paints
Oleo-resin yielding broad-leaved species		
Boswellia serrata	Guggal/salai	Rheumatism or nervous diseases
Dipterocarpus turbinatus (gurjan)	Gurjan oil	Lithographic ink, anti-corrosive agent, timber preservation and medicinal use
Kingiodendron pinnatum (piney)	Piney oil	Wood varnish, medicinal use
Gum-resin yielding broad-leaved species		
Garcinia morella	Gamboge	Varnishes, medicinal use as cathartic
Commiphora mukul	-	Perfumery fixative agent, medicinal use

- Leaves of *Butea monosperma* are used for making **platters and cup**. While leaves of *Diospyros melanoxylon* are used for bidi making.
- **Saponin** (substitute of soap) is obtained from *Sapindus mukorossi*. Beads seeds for making necklaces are obtained from *Abrus precatorius, Elaeocarpus sphaericus* and *Putranjiva roxburghii.*
- **Drugs** are obtained from roots, leaves, barks, flower, fruit and seeds of a tree.

Root drugs	Bark drugs	Flower/fruit/seed drugs	Leaf drugs
Podophyllum hexandrum	*Cinchona hybrida*	*Artemisia species*	*Ephedra gerardiana*
Rauwolfia serpentine	*Holarrhena antidysentrica*	*Strychnos nux-vomica*	*Vitex peduncularis*
Dioscorea deltoidea	*Soymida febrifuga*	*Aegle marmelos*	*Gaultheria fragrantissima*
Acorous calamus		*Ricinus communis*	*Mentha species*
Asparagus adscendens		*Cassia fistula*	*Ocimum killimandscharicum*
Rheum emodi		*Chenopodium species*	*Azadirachta indica*
Picrorhiza kurrooa		*Plantago ovata*	*Cannabis sativa*
Valerian wallichii		*Pyrethrum species*	*Atropa acuminata*
Glycyrhiza glabra		*Caesalpinia erista*	*Datura species*
Berberis aristata		*Tamarindus indica*	*Swertia chirata*
Aconitum species			
Saussuria lappa			

- Table showing gross structure, minute structure, physical and mechanical properties of wood.

Gross structure	Minute structure	Physical properties	Mechanical properties
Pith	Vessels or pores	Colour	Hardness
Sapwood	Fibres	Lustre	Flexibility
Hardwood	Tracheids	Odour	Fissibility
Bark	Tylosis	Weight	Elasticity
Spring wood/early wood	Ring-porous wood	Grain	Strength
Autumn/late wood	Diffuse-porous wood	Texture	Aptitude for being worked
Growth rings	Pith flecks	Figure	
	Gummy deposit		
	Parenchyma or soft tissue		
	Wood rays		
	Ripple marks		
	Gum & resin duct		

- **Combustibility** of wood refers to capacity of wood to catch fire and continue to burn until it is consumed. Combustibility of wood signifies readiness.
- The quantity of heat emitted by a given weight of wood during combustion process is called **heating power or calorific value of wood**. The two element which burn in wood are carbon (C) and hydrogen (H).
- **Calorific** value indicates the number of grams of water to raise 1°C temperature of 1 gram wood when completely burnt. Calorific value of some important woods are given below:

Species	Common name	Calorific value
Cedrus deodar	Deodar	5294
Acacia nilotica	Babul	4870
Casuarina equisetifolia	Casuarina	4950
Anogeissus latifolia	Axlewood	4948
Shorea robusta	Sal	5264
Tamarindus indica	Imli	4939
Quercus species	Oak	3990

- **Teakwood** is not a suitable fuel wood for cooking due to its smokiness. Wood which have more resinous substances are good as fuel.
- **Wood defects** refers to "Any of the various imperfections that can be observed in lumber and wood products, such as checks, splits, knots, crooks, bowed wood, machine burns, sap-stain, blue-stain, cupped wood, twists, and wormholes"
- The various defects may be grouped under the following main heads:

Defects due to	
Abnormal growth	**Rupture of tissue**
Waviness	Checks
Twisted fibre	Split
Knots- Live or tight knots - Dead or loose knots	Shakes- Heart & star shakes - Radial shakes -Cup & ring shakes
Burr	
Constriction due to climber	
Interior bark	

- A **knot** is a part of a branch embedded in wood. A portion of branch which is living at the time of its inclusion and establishes a fibrous connection with the surrounding wood is called **live knot**. Live knots do not separate when wood dries. Live knots are not serious defect.
- **Dead knots** do not have a fibrous connection with thc main stem and reducc the strength of wood. These knots are detached during seasoning.
- Defect caused by fungal attack and decay in wood is called as **unsoundness**. It is also called fungal decomposition of wood.
- The most serious fungal defects of wood are **rots and stains**.
- When the wood grain twisted spirally making an angle 40° with vertical axis is called **twisted fibre**. Usually wood grain run straight with vertical axis. Twisted fibre most commonly found in *Hardwickia binata, Boswellia serrata, Casuarina equisetifolia* and *Pinus roxburghii*.
- When wood fibres shows a wavy appearance instead of true vertical straight line with main axis it is called **waviness**.
- **Burr** is a complex knot formed at the point where dormant bud shows abnormal growth without developed into branches resulting formation of concentrated mass around bud.
- Separation of fibres forming a cracks or fissure in the piece of wood not extending to other face or end of wood piece is called **checks** commonly it occurs in converted timber.
- Separation of fibres forming a cracks or fissure in the piece of wood and extending to other face or end of wood piece is called **split**.
- A crack starting from the pith region and extending radially outwards is called **heart or star shakes**. When more than one radially cracks formed towards the pith called **compound heart shakes or star shakes**. These are common in old trees.
- Cracks which start from the periphery and extend radially inward towards the centre is called **radial shakes**. These are formed due to rapid contraction of outer growth zone as compare to inner growth zone.
- When the cracks follows the direction of annual rings due to shrinkage of central tissues and looks like as cup it is called **cup and rings-shakes**. These are mechanical defects.
- Rindgalls are mechanical defects in timber due local wounds by wild animals or wild fire.

- Two main defects in the wood due to fungal attack are **stain** and **decay**. In stain defects fungi attacks only sapwood and responsible for sapwood discolouration. While in decay defects fungi attacks on sapwood as well as heartwood and is responsible for wood decomposition. Wood decomposition by fungi is called **decay**.
- The **wood rotting fungus** draws its food from the cell wall material of wood through hyphae and decay both heartwood as well as sapwood.
- Table showing defects caused by insects in timber:

Insects	Insect attacks on	Preventive measures
Powder-post borer	Sapwood	Steam sterilization & chemical soaking or spraying of greenwood
Bark beetles	Bark of felled tree	Debarking & quick seasoning of felled wood and chemical treatment
Flatheaded borers	Bark & sapwood	Debarking of felled wood soon
Longhorn borers	Bark, sapwood & heartwood	Spraying wood with chemical
Carpenter bees	Hole in timber	Spraying wood with chemical

- There are about 200 species of termites in the country. Main food of termites is **cellulose**.
- Marine borer such as *Limnoria, Chelura* and *Sphaeroma* attack on wood surface and reduce strength of wood. *Tredo, Bankia* and *Martesia* genera are also known as **shipworms**.
- Unlignifierd area in primary cell wall is knows as **pits**. These are responsible for intercommunication between two adjacent cells.
- Those features of wood which can be seen with naked eyes or with an ordinary lens is called **gross features**. While those features of wood which can see with compound microscope or more clear lens is called **minute features of wood**.
- Central portion of the wood is called **pith**. While bark is a diagnostic value in identification of trees and wood. Odour is more pronounced in heartwood than in sapwood.
- The **colour** in wood is due to infiltration of chemical products. **Lustre** is the ability of cell wall to reflect the light. This property is not seen by all timbers. *Picea smithiana* (Spruce) and *Boxus sempervirens* (boxwood) have characteristics lustre.

- The **density** of any material is defined as weight of unit volume of the material and expressed as kg/m^3. **Specific gravity** is defined as ratio of the weight of material to the weight of an equal volume of water at 4°C. Specific gravity of water is 1 gm/cm^3.
- Moisture content in **green timber** vary from **50-200%**. While in air **dry timbers** have **8-15%** moisture content. **Softwoods** loose relatively more weight as compare to hard woods during air drying.
- Table showing classification of timbers based on the air dry weight:

Timber	**Density (kg/m^3)**	**Species examples**
Very light	300	Suji (*Cryptomeria Japonica*)
Light	301-450	Silver Fir, Semal
Moderately heavy	451-600	Siris, Deodar, Walnut
Heavy	601-800	Teak, Mulberry
Very heavy	801-950	Babul, Sisso, Dhonkra, Sal
Extremely heavy	951 or above	Oaks, Rohan (*Soymida febrifuga*)

- **Grain** is the alignment of cells or direction of fibres regards to vertical axis of the stem. Grain may be straight, irregular, diagonal, spiral, interlocked and curly or wavy. Interlocked fibres does not reduce timber strength but other types of grains reduce timber strength.
- **Texture** of wood refers to arrangement of cells according their size in unit volume of wood.
- **Figures** refers to distinctive pattern produced on longitudinal surface of timber due to alignment of tissues and direction of the grain.
- Table showing classification of timbers according to various property:

According hardness	
Classified wood	**Examples**
Extremely hard	Kusum, *Mesua ferrea*, sal babul, rosewood
Very hard	*Terminalia alata*, Sissoo, *Morus alba*
Hard	*Chukrasia tabularis, Tectona grandis*
Moderately soft	*Mangifera indica*
Soft	*Lannea coromandelica*, deodar, silver fir
Very soft	*Boswellia serrata*, chirpine, spruce
Extremely soft	Semal, *Cryptomeria japonica*

According elasticity	
Classified wood	**Examples**
Very elastic	*Grewia tiliaefolia, Anogeissus latifolia, Mesua ferrea*
Elastic	*Chloroxylon swietenia*, sal, babul, sissoo, rosewood
Moderately elastic	*Mangifera indica, Toona ciliata, Tectona grandis*
According fissility	
Classified wood	**Examples**
Extremely fissile	*Tectona grandis*
Easily fissile	*Acacia catechu, Adina cordifolia, Toona ciliata*
Moderately fissile	*Bombax ceiba, Anogeissus latifolia*
Difficult to split	*Dalbergia sissoo & Shorea robusta*
Very difficult to split	*Dalbergia latifolia, Ougenia oojeinensis*
According strength	
Classified wood	**Examples**
Exceptionally strong	*Mesua ferrea, Shorea robusta*
Extremely strong	*Acacia nilotica, Grewia tiliaefolia, Terminalia alata*
Very strong	*Tectona grandis, Anogeissus latifolia*
Strong	*Dalbergia sissoo, Dalbergia latifolia, Albizia lebbeck*
Moderately strong	*Adina cordifolia, Cedrus deodara, Chukrasia tabularis*
Somewhat weak	*Abies pindrew, Pinus roxburghii*
Weak	*Picea smithiana, Lannea coromandelica,*
Very weak	*Hymenodicton excelsum, Bombax ceiba, C. japonica*

- **Flexibility** is the capacity of wood to bend out of shape freely without rupture.
- **Elasticity** is the capacity of wood to regain its original shape after the stress has produced deformation.
- **Fissility** or **fissibility** is the capacity of wood to split along the fibre during conversion of wood with axe or a wedge.
- **Wood seasoning** refers the process of removal of excess moisture present in timber in its green state.
- **Horizontal** and **vertical stacking** is used in air seasoning. One in nine method, close rib method and open rib methods are kind of horizontal stacking (30-45 cm above ground level).

- Table showing time required for wood seasoning:

Class of wood	Seasoning season	Wood thickness (in cm)	Drying period (month)
Highly refractory	Ending rainy season	2.5 5.0	6 12
Moderately refractory	Ending winter (Feb-Mar)	5.0 5.0	4 8-10
Non refractory	Hot dry summer	2.5 5.0	2 4
Railway sleepers	Ending rainy season	-	12-24

- **Wood prescrvation** is defined as the process of improving wood's natural durability by treatment with chemicals that are toxic to insects, fungi and other decaying agents.
- **Wood preservatives** are the chemicals that are applied to woods to make it resistant to attack by different decay agents.
- Wood preservatives are classified into three categories viz. oil types, water soluble types and organic solvent types.
- Table showing characteristics of wood preservatives:

Types	Examples	Effective against	Remarks
Oil types	Creosote, coal tar	Borer, marine organisms, termites.	Leaching resistance, increase wood inflammability, not suitable for indoor works.
Water soluble types 1.Non-fixed 2. Fixed	Borax, Boric acid, Sodium PCP, Cu-sulphate, $HgCl_2$, Na fluorides, $ZnSO_4$ etc. CCA, CCB, CCF, CCP, Chromated $ZnCl_2$, Zn Meta Arsinate (ZMA), Ammonical Cu-Chromate (ACA), Acid Cu-Chromate (ACC)	Various decay, easily washed, applied by dip diffusion. Leaching resistant, non-corrosive	Not suitable for outdoor works, poisonous, cause swelling in woods.
Organic solvent type	PCP, Lindane, copper napthalate, copper quinolate.	Fungi, other microbes	Very expansive, increase wood inflammability

- **Free water** refers to the water present in the cavity of the wood cell. **Bound water** refers to the water held within the cell walls by hydrogen bonding.
- There are two methods of seasoning viz. air seasoning and kiln seasoning.
- **Air seasoning** refers to the drying of wood with the help of air and sunlight.
- **Kiln** is air tight, thermal proof, enclosed chamber with provision for controlling temperature, air circulation and humidity.
- Table showing constituents of water soluble preservatives by weight:

Types	As_2O_5	$CuSO_4$	Na/K Cr_2O_7	Acetic acid	Boric acid
ASCU	1% by wt.	3%	4%	–	–
ASCU-Boric	–	3%	4%		1.5%
Celcure	–	5.5%	5.5 (Na)	0.2%	–

- **Vertical stacking** is used only for the rapid surface drying of certain species of non-refractory woods.
- Wood preservation on scientific and modern basis was introduced in India by **Sir Ralph Pearson** of the Indian Forest Services in the year 1908.
- Osmose process and hot & cold process are applied in **soaking method**.
- Table showing wood preservation processes with their characteristics:

Non pressure method	
Steeping	Useful for water soluble preservatives
Soaking	Useful for dry as well as green woods
Diffusion	For green timber with water soluble preservatives
Sap displacement method	Green round timber & bamboo with water soluble preservatives
Heating and cooling	Wood kept for 3 hrs at 100°C.
Pressure method	
Full cell process	Pressure applied at 3.50 to 12 kg/cm^2
Empty cell process	Maximum penetration is obtained with min. absorption
Lowry process	No need to create initial vacuum, use for thatches & ropes
Rueping process	Pressure applied at 5.0 to 12 kg/cm^2

- **Boulton** process of preservation combines both seasoning and preservation.
- Water transportation, land transportation and overhead transportation are the method of transportation of woods.
- Water transportation is extensively used in the coniferous forest of N-W Himalayan.
- Water transportation includes floating, rafting & boom and wetslides.
- Softwoods are transported through floating methods of transportation.
- Telescopic floating and river floating or free floating or sweeping are the subtypes of floating method of wood transportation.
- **Pathru** is used in telescopic floating of woods. This methods is only used in the hilly forest of N-W Himalayan.
- The average distance covered daily in the floating operations is 2-3 km. This methods is used to transport the woods for a distance upto 100-200 km.
- A quantity of woods firmly bound together is termed as raft section and the process is called **rafting**.
- The average distance covered in the **rafting operations** is 40-60 km per day. This method is mostly used in winter seasons.
- The **rafts** are classified into log raft, sawn timber rafts and bamboo rafts.
- **Logs rafts** contains 6-10 sections and each section contains of 16-20 logs. The average logs in a log raft is 96-200 logs.
- Bamboos are tied into a bundle of 50-100 culms. The average number of bamboos carried in a raft is 10,000.
- Boom is a Dutch word which means **floating obstruction**.
- Single-arm or one-way boom and two-way or V-shape booms are the types of **booms**.
- **Single-arm or one-way boom** is attached permanently at it upper ends and guy ropes always attached at right angle (90°) to the boom.
- The arms of **two-way or V-shape boom** incline at an angle 45° at the apex on the downstream side. While apex of boom is kept at point of maximum velocity.
- Boom proper and log line or **line dori** are used in Single-arm or one-way boom.
- **Wetslides** are used in the inaccessible hilly areas to transport woods.

- **Composite** wood is the general term for building bonded product consisting either wholly of natural woods or of wood in combination with metals, plastics etc.
- Plywood is glued wood made up of veneers in such a manner that the grain of each veneer is at right angles to the adjacent veneer in the assembly. The outer faces are called as **faces** and centre one is called as **core**. The numbers of veneers is always odd in number.
- In a panels having more than three plies the layers between the *core* and the **face** or **back** are called **cross bands**. The simplest structure is of the three plies.
- The maximum length of logs used in plywood is about **255-280 cm**.
- The pressing of glued coated veneers in plywood is done at a pressure range of **7-18 kg/cm²**. And moisture content in plywood is brought upto **12%**.
- The grains direction in laminated wood is parallel to the adjacent veneer. While in core boards it is perpendicular to the adjacent veneer.
- The glue laminated wood is called as **glulam**. Lamina refers to the thin board of veneers.
- A core board having width of 7.5 cm is called batten board. A core board not more than 2.5 cm wide is called **block-board**. A core board having width less than 7.0 mm is called **lamin board**.
- The thin faces of sandwich board called are as **skins**.
- **Fibre boards** are classified into hard board, semi hardboards, medium hardboards and soft hardboards.
- Common particle boards are chip-boards, flake boards and shaving boards.
- Improved wood is also known as **modified wood**. Improved wood may be classified into impregnated wood, heat stabilized wood, heat stabilized compressed wood, compressed wood, compregnated wood and chemically modified wood.
- The term **impregnated wood** is a collective name for woods impregnated with all possible impregnates. Common used impregnates are wax, paraffin, resin and oils.
- Veneers treated with a thermosetting fibre penetrating resin is called as **impreg**.
- **Heat stabilized wood** is treated at high temperature (260°C-315°C) for a few minutes under non-oxidising conditions.

- **Compressed wood** can be made either from solid wood or laminate wood by applying pressure from 90-140 kg/cm^2.
- **Compregnated wood** is highly resistance to mild acids, alcohols and many other solvents.
- **Compregnated wood** has 10-20 times more hardness and has increased wear resistance as compared to normal wood. It is an excellent materials for **aeroplane propeller blades**.

CHAPTER 6

Plant and Wood Anatomy

- The branch of botany that deals with the study of internal structures and organization of plants or plant organs is called **Plant Anatomy**.
- **N. Grew** is known as father of plant anatomy and K. A. Chaudhary is father of Indian plant anatomy.
- A group of cells which is similar or dissimilar in shape, having a common origin and usually perform a common function is called a **tissue**. Term tissue was coined by N. Grew.
- Plastids are absent in meristems.
- The meristems which occurs at the tips of roots and shoots and produce primary tissues are called **apical meristems**. These are responsible for increase in the length of plants organs.
- The meristems which occurs between mature tissue is called **intercalary meristem**.
- Intra fascicular or fascicular cambium occurs inside the vascular bundles.

Theory	Given by
Apical cell theory	Karl Nageli & Hofmeister
Histogen theory	Hanstein
Tunica corpus theory	Schmidt
Mantle core theory	Popham & Chan
Newman's theory	Newman
Korper-Kappe theory	Schuepp

- Permanent tissues cells are composed of cells (live or dead) which have lost the power of division temporarily or permanently.

- Simple tissues are three types:

Types	Founder	Function
Parenchyma	Grew	Storage of food
Collenchyma	Schleiden	Provide mechanical strength
Sclerenchyma	Mettenius	Main mechanical tissue

- **Aerenchyma** provide buoyancy to hydrophytes. Parenchyma is a universal tissue in plants.
- **Xylary fibres** or wood fibres are hard fibres and are obtained from xylem.
- Bast fibres or extra xylary fibres are also known as commercial fibres and obtained from phloem and pericycle of plants. Cotton fibres are composed of cellulose.
- Xylem and phloem collectively are called as vascular bundles.
- Complex tissues are made of more than one types of cells and are of two types viz. xylem and phloem.
- Table showing elements and functions of xylem and phloem:

Parameters	Xylem	Phloem
Term	Nageli	Nageli
Elements	4 elements: Tracheids – Dead elements (DE) Vessels – Dead elements Xylem fibres – Dead elements Xylem parenchyma- living elements (LE)	4 elements: Sieve cells/sieve tubes-(LE) Companion cells-(LE) Phloem fibres-(DE) Phloem parenchyma-(LE)
Function	Conduct water & minerals salts	Conduction of food materials

- **Tracheids** are primitive conducting elements of xylem in plants.
- The maximum bordered piths are found in the Tracheids of **Gymnosperms**.
- Vessels work as a pipe line during conduction of water in plants.
- Tracheids and vessels are collectively known as water conducting elements or **Hadrom**. Term hadrom and totipotency was given by **Haberlandt**.
- Formation of whole individual from a single cell is called **totipotency**.
- Sieve elements was discovered by **Hartig**. Companion cells are found only in Angiosperms.

- The conducting elements of phloem is called **Leptom**. Term leptom was given by Haberlandt.
- When the xylem and phloem are present separately on different radii in alternate manner it is called **radial vascular bundles**. These are also called as exarch. Examples- most of the roots.
- When the xylem and phloem are present on same radius it is called **conjoint vascular bundles**. These are also called as endarch. Examples- most of the stems.
- When cambium is absent between the xylem and phloem it is called **closed vascular bundles.** These are found in monocotyledon stems.
- When cambium is present between the xylem and phloem it is called **open vascular bundles**. These are found in dicotyledons and gymnosperm.
- Conjoint bicollateral and open vascular bundles are found in stems of family cucurbitaceae.
- The whole central mass of vascular tissue with or without pith surrounded by endodermis is called **stele**. This concept was given by **Van Tieghem** and **Duoliot**.
- Protostele is most primitive type of stele while atactostele is most advanced type of stele.
- In dicot stem cortex is divided into hypodermis, general cortex and endodermis.
- **Pith** is well developed in monocot roots and dicot stems.
- **Casparian strips** are found in endodermis of dicot roots.
- Pith is less developed or absent dicot roots and monocot stems.
- The central region of secondary xylem which is dark in colour is called **heart wood or Duramen**. The function of heart wood is to provide mechanical strength to the stem.
- Tables showing classifications of woods:

Wood types	Contains	Found in
Manoxylic/soft woods	More living parenchyma	Gymnosperm
Pycnoxylic/hard woods	Less amount of living parenchyma	Pinus, sissoo, teak
Non porous/soft woods	Vessels are absent	Gymnosperm
Porous/hard woods	Vessels are present	–
Ring porous woods	Vessels are arranged in ring form	Temperate tree
Diffused porous woods	Vessels are arranged a systematically	Tropical tree

- Normally secondary growth takes place in roots and stems of dicot & Gymnosperms.
- Due to lack of cambium in monocot, secondary growth is absent. But exceptionally secondary growth takes place in some monocotyledons. Examples- Yucca, Dracaena, Coconut, Agave and Smilax.
- The wood formed during the spring season called is **spring wood** or **early wood**. Spring wood is lighter in colour and exhibits low density.
- The wood formed during the winter season is called **autumn wood** or **late wood**. Autumn wood is darker in colour and exhibits high density.
- More distinct annual rings are formed in temperate plants while distinct annual rings are not formed in tropical plants.
- The peripheral region of secondary xylem which is light in colour is called **sap wood** or **Alburnum**. The function of sap wood is conduction of water and minerals.
- Non-porous wood is also called as **homoxylous wood** and porous wood also called as **heteroxylous wood**.
- Phellem, phellogen and phelloderm are collectively called as **periderm**.
- Lenticels are formed by the activity of phellogen.
- Lenticels are mainly found on woody stems and never found on leaves.
- All the tissues situated outside the cork cambium is called **bark**.
- Outer layer of bark is called **rhytidome** and inner layer of bark is called secondary phloem.
- Continuous bark of equal thickening is called **ring bark**. In ring bark cambium is continuous. Example- Bhojpatra (Betula) and eucalyptus.
- Discontinuous bark of unequal thickening is called **scaly bark**. In scaly bark cambium is not continuous. Example- Azadirachta, Mango and Tamarindus etc.
- A ring of autumn wood and a ring of spring wood are collectively known as **annual ring**. The number of annual rings, formed in a tree gives the idea of the age of the tree. The study of determination of age of the plant by this techniques is called **Dendrochronology**.
- Study of the wood is called **xylotomy**. The wood is a secondary xylem.

CHAPTER 7

Forest Mensuration

- Mensuration is derived from the Latin word *mensura* meaning **measure**.
- The term *Pramana* meaning 'measure' was classified into four types:
  - *mana-* measure of capacity or volume
  - *tula mana-* measure of weight
  - *avamana-* linear measure
  - *kala pramana-* measure of time
- Kautilya has derived *gorut* (12 sq. km) as the area unit of forest.
- **Forest mensuration** is the branch of the forestry which deals with the determination of dimensions (diameter, height, volume etc.) form, age and increment of single trees, stands or whole woods, either standing or after felling.
- **Volume** is the most important measurement from the point of view of forest management.
- Forest mensuration have following objects:
  - Basis for sale
  - Basis of management
  - Measurement for research
  - Measurement of planning
- The Standards of Weights and Measure Act was formed in 1956.
- Metric system or C.G.S system was introduced in India in 1962.

Parameter	C.G.S or Metric Unit	F.P.S or British Unit
Length	Metre	Foot
Area	Square centimetre	Square foot
Volume	Cubic metre	Cubic foot

- There are two system commonly used for measurement-
  1. British system or F.P.S system
  2. French or Metric system or C.G.S system
- In forest mensuration, cubic metre (m^3) is used for tree volume measurement.
- Metric system is most commonly used system in India.

Parameter	Measurement unit (in metric system)
Tree height crown height, crown width, crown length	Metre, correct to first place to decimal
Stump height or seedling height	Nearest centimetre
Diameter	Centimetre
Girth	Metre and centimetre

- **Laser** is Light Amplification by Stimulated Emission of Radiation.
- The circumference or perimeter of a circle is = $2\pi r$
- **Breast-Height** is defined as almost universal adopted standard height for measuring girth, diameter and basal area of standing trees.
- In India Breast-Height is taken as **1.37 m (4′ 6″)** while in Europe and United Kingdom it is taken as **1.3 meter (4′ 3″)**.
- Standard rules for the measurement of the Breast-Height are given below:
  1. Breast-Height should be marked with a measuring stick on a standing trees at 1.37 meter above the ground level.
  2. The Breast-Height point should be marked by intersecting vertical and horizontal lines 12 cm long, painted with white paint. This is referred to as **cross-mark**.
  3. On slopping ground diameter at Breast-Height should be measured on the uphill side.
  4. Diameter of leaned tree at flat ground should be measured along tree leaned side not vertically.
  5. The **Diameter at Breast-Height** (d.b.h) should not be measured at 1.37 m if the tree is abnormal at the level. Breast-Height mark should be shifted up or down as little as possible.
  6. When the tree is forked above the Breast-Height, it is counted as one tree. When the tree is forked below the Breast-Height, each fork should be counted as separate tree.

7. For buttress tree Breast-Height should be **measured** from the ending point of buttress formation.
8. Moss, creepers, lichens and loose bark found on the tree stem must be removed before measuring the diameter or girth over bark.
9. Diameter measurements should be recorded in centimetres and to the nearest multiple of two millimetres.

- Wooden scale, calliper and tape are used for the measurement of tree diameter or girth.

Instruments	Used for measure	Error
Wooden scale	Diameter of stump or logs	–
Calliper	Diameter & girth of standing tree	Negative & positive
Tape	Diameter & girth of standing & felled tree	Positive

- D.O.B can be converted into D.U.B by deducting twice bark thickness.
- G.O.B can be converted into G.U.B by the formula:

$$\text{Girth under bark } (g) = g' - 2\pi t$$

g = girth under bark, g' = girth over bark, t = bark thickness

- The thickness of bark can be measured by **Swedish bark gauge**.
- **Seth and Lohani** have prepared bark thickness table for Picea smithiana (spruce).
- The usual diameter classes used in India are 2 cm, 5 cm and 10 cm, as shown in the table.

Diameter class (in cm)	Tree attain maturity
2	30 cm d.b.h
5	30-50 cm d.b.h
10	50 cm and above 50 cm

- The usual girth classes used in India are 5 cm, 15 cm and 30 cm.

Girth class	Tree size
05	Small-sized
15	Medium-sized
30	Large-sized

Diameter class	Colour
0-20	Blank
20-30	Green
30-40	Red
40-50	Yellow
50-60	Black
60-70	White
70-80 or above	Blue

- The ratio between the diameter (d) and girth (g) of a circle is $1/\pi$ or 0.3182. But this ratio varies from species to species.
- **Dendrometer** is used to measure the diameter of a standing tree at greater height.
- **Spiegel Relaskop** is used to measure upper stem diameter. For diameter measurement Band 1 and 4 narrow bands are used.
- **Wheeler Penta Prism Calliper** is used to measure stem diameter accurately at any point on the tree stem from any convenient distance. No base line distance is required. It gives diameters correctly upto ±2 millimetre.
- Total height of a standing tree is the vertical straight line distance from the tip of the leading shoot to the ground level.
- Bole height is the distance between ground level and crown point.
- Commercial bole height is the height of bole that is usually fit for utilization as timber.
- Height of Standard timber bole is the height of the bole from the ground level upto the point where average diameter over bark is 20 cm.
- Height of the stump above ground is called **stump height**.
- **Crown length** is the vertical measurement of the crown of a tree from the tip to the point half way between the lowest green branches forming green crown all round and the lowest green branch on the bole.
- **Crown height** is the height of the crown as measured vertically from the ground level to the point half way between the lowest green branch and green branches forming green crown all round.

- Crown width is defined as the maximum spread of the crown along its widest diameter.
- Methods of height measurements are ocular, non-instrumental and instrumental methods.
- **Shadow method**- In this non-instrumental method a pole of convenient length is fixed upright in the ground and its height above the ground is measured. This method can be applied only on sunny days.
- **Single pole method**- In this non-instrumental method observer holds a pole of about 1.5 m length vertically at arm's length in one hand in such a way that the pole portion above the hand is equal in length to the distance of the pole from the eye.

Height measuring instrument based on	
Similar triangle	**Trigonometric principle**
Christen's Hypsometer	Brandis Hypsometer
Smythie's Hypsometer	Abney's level
Improved callipers	Hega Altimeter
	Topographical Abney's level
	Relaskop
	Tele Relaskop
	Bar & Stroud Dendrometer
	Blume-Leiss Hypsometer

- Height class is defined as **n f** the intervals into which the range of tree heights is divided for classification and use. In India 1, 3 and 5 metre height classes are generally used.

Height class (in metre)	**Tree attained a maturity (mm)**
1	<15
3	15-25
5	> 25

- Form is defined as the rate of taper of a log or stem.
- Taper is decrease in the diameter of a stem of a tree of a log form base upwards.

Tree portion	Similar to	Equation
Basal portion	Frustum of a neiloid	$Y = Kx^3$
Middle portion	Frustum of a paraboloid	$Y = Kx$
Top portion	Like cone	$Y = Kx^2$

- According to **Metzger' theory**, the tree stem should be considered as a cantilever beam of uniform size against the bending force of the wind. This is also called as **Girder theory**.

$$d^3 = \frac{32 \times W \times F \times L}{p \times S}$$

- According to Metzger' theory, the tree stem must have the shape of a cubic paraboloid.
- Form factor is the ratio between the volume of a tree to the product of basal area and height.
- Artificial form factor also called as breast-height form factor. For this form factor, the basal area is measured at the Brest Height and volume refers to the whole tree both above and below the measurement point.
- In **absolute form factor basal** area is measured at any convenient height and volume is measured above the measurement point.
- In **normal (true) form factor basal** area is measured at a constant proportion of the total height of the tree and volume is measured as whole of the tree above the ground level.
- Artificial form factor is most commonly used form factor in India.
- The formula of form factor is given below:

$$F = \frac{V}{S \times h}$$

F = form factor, V = tree volume (m^3), S = basal area, h = tree height.

- Form factor is used to study the growth law and to estimate the volume of standing tree.
- Form factors are classified into three kinds:
  1. Tree form factor
  2. Stem timber form factor
  3. Stem small wood form factor

- **Form height** is defined as the product of form factor and total height of the tree. The formula of form height is given as:

$$Fh = \frac{V}{S}$$

Where Fh = form height

- The concept of form quotient was given by Austrian forester **A. Schiffel**. It is the ratio between the mid-diameter and d.b.h. This is also known as normal form quotient

$$\text{Form quotient} = \frac{\text{mid-diameter}}{\text{d.b.h}}$$

- Swedish forester **Tor Jonson** modified the form quotient. He defined the form quotient as the ratio of diameter or girth of a stem at one half its height above the breast-height to the diameter or girth at breast-height. This is also known as **absolute form quotient**.
- Form class is defined as one of the intervals in which the range of form quotients of trees is divided for classification and use.
- Form point is defined as the point in the crown at which wind pressure is estimated to be centred.
- Form point ratio is defined as the relationship of the height of the form point above ground level to the total height of the tree.
- Taper table is used for the study of tree forms.
- **Hojer and Behre** formula are used to determine diameter quotient.
- The ratio of the diameter (d) of a stem at any given height to its breast-height diameter (d.b.h) is called **diameter quotient**.

Hojer's formula $\frac{d}{\text{d.b.h}} = C \log \frac{c+l}{c}$

c & C = constant for each class, l = distance from tree top at diameter measurement point.

Behre's formula $\frac{d}{\text{d.b.h}} = \frac{l}{a+bl}$

- The area of the bole surface of a tree and log can be calculated by the formula:

$$S = g.l \text{ or } \frac{g_1 + g_2}{2} \times l$$

S - bole surface area, l - length of tree/log, g_1 & g_2 - girth at thicker & thinner end.

- Crown cover is defined as the horizontal projection on the ground of the tree crown.

Name of body	Area formula
Circle	$A = \pi r^2$ or $\frac{\pi d^2}{4}$ where, r-radius & d-diameter
Square	a^2 a = side of square
Rectangle	Breadth (b) × length (l)
Triangle	$\frac{\text{base} \times \text{height}}{2}$
Circular section	$\pi r^2 \times \frac{\theta}{360}$
Ellipse	πab a & b are semi major and minor axis
Parabola	$\frac{2}{3} \times ld$ l-length of base, d-depth of parabola

- The following table gives the formulae for calculation of volume logs:

Form of solid	Volume of full solid	Vol. of a frustum of solid	Remarks
Cylinder	Sl	Sl	
Paraboloid	Sl/2	$\frac{s_1 + s_2}{2} \times l$	Smalain's formula
		$S_m \times l$	Huber's formula
Cone	Sl/3	$\frac{(s_1 + s_2) + \sqrt{(s_1 \times s_2)}}{3} \times l$	
Neiloid	Sl/4	$\frac{(s_1 + 4Sm + s_2)}{6} \times l$	Newton's formula

- Newton's formula is used to calculate the error in the volume calculated by **Smalian or Huber formula.**
- Smalain's formula requires the end cross-sections of the log for volume calculation.
- Huber's formula requires the mid-cross-sections of the log for volume calculation.
- The difference between volumes obtained by Smalain's and Newton's formula is:

$$\Delta = \frac{l}{3} \times (s_1 + s_2 - 2Sm)$$

- The difference between volumes obtained by Huber's and Newton's formula is:

$$\Delta = \frac{1}{3} \times (s_1 + s_2 - 2Sm)$$

- In case of cylinder and paraboloid there is no difference between volumes obtained by Smalain's and Newton's formula. But in case of cone and **neiloid** the error is **positive.**
- In case of cylinder and paraboloid there is no difference between volumes obtained by Huber's and Newton's formula. But in case of cone and neiloid the error is **negative.**
- In India, volumes of logs are calculated by using quarter girth formula. It is also known as **Hoppus's rule in Britain**.

$$\text{Volume of log} = \left(\frac{g}{4}\right)^2 \times l$$

Where, g = log girth, l = log length.

- The volume derived from this formula is lesser than true volume because here pie is assumed equal to 4.
- This formula gives only **78.5%** of the true volume.
- The space occupied by a stack as distinct from the cubic contents of the wood itself i. e. solid volume is called **stacked volume.**
- Volume of solid firewood can be calculate by using xylometric and specific gravity method.
- The measurement of pulp wood is done in stacks and conversion factor are used to convert stacked volume into the solid volume.

Stacks type	Value of conversion factor
Normal packed stacks	0.50 to 0.66
Loosely packed stacks	0.30
Well (tight) packed stacks	0.80

- Volume of standing tree can be measure by the following methods:
    1. Ocular estimate
    2. Partly ocular and partly by measurement
    3. Direct measurement
    4. Indirect measurements
- **Spiegel Relaskop, Tele Relaskop, Wheeler Penta Prism callipers** and **Barr**, and **Stroud Dendrometer** are used in indirect measurement methods.
- Volume table is defined as a table showing a specie's average contents of trees, logs or sawn timber for one or more given dimensions.
- The volume of a tree depends mainly upon three variables viz., diameter, height and form. Diameter at breast-height is the most important of the tree variables.
- Classification of volume tables are shown in table given below:

Classification on the basis of variables	
Volume table based on one variable	Diameter
Volume table based two variables	Diameter and height
Volume table based on three variables	Diameter, height and form quotient
Classification on the basis of scope of application	
General volume table	Diameter and height (two variables)
Regional volume table	–
Local volume table	Diameter (one variable)
Classification on the basis of outturn kind	
Standard volume table	For standard round timber estimation
Commercial volume table	For commercial round timber estimation
Sawn outturn table	For sawn timber estimation
Assortment table	Round timber volume in various form
Sawn assortment outturn table	Sawn outturn in the form of standardized sized pieces

- Volume table based on one variable is also called **local volume table**.
- Volume table based on three variables have not been prepared in India.
- General volume tables are used for the deriving local volume tables. And most commonly used in India.
- Standard volume table and Commercial volume table are the special cases of assortment table.
- The volume tables are prepared by the following methods:
  a. Graphical method
  b. Regression equations method or method of least squares fit
  c. Alignment chart method
- The general **tariff tables** of Europe consist of a series of 51 related volume tables.
- The portion of the tree stem or log which is unmerchantable is called **cull**.
- Volume table is used to estimate the volume of standing trees.
- Density, moisture content, bark & foreign material affect the weight of the wood.
- Mass per unit volume of a substance is called **density**. While the ratio of density of the substance to density of water is called as relative density or specific gravity.
- **Limaye & Sen** reported the specific gravity of certain Indian tree species based on oven-dry and green volume. Bark has lower density than the wood.
- The moisture content when free water has been evaporated leaving only absorbed or bound water is called **fibre saturation point**.
- Conifer heartwood have lower moisture percentage as compare to sapwood.
- Determination of the age of single standing tree can be estimated by using following methods:
  1. From existing record.
  2. From general appearance.
  3. By the number of annual shoots or whorls of branches- in case **Bombax ceiba** (Semal).
  4. By means of **Pressler's increment borer**.

- Age of a tree can be determined by counting annual ring with the help of Pressler's increment borer. This methods is suitable only for species showing distinct annual rings.
- **Boring** is done 30 cm above ground in case of smaller tree and at breast-height for bigger tree.
- The age of standing tree without counting annual ring can be calculated by taking three periodic mean. The age of a tree can be found by the following formula:

$$\text{Age} = \frac{1}{P_1 \times s} \quad \text{where, } s = \frac{\log P_1 - \log P_2}{\log d_2 - \log d_1}$$

- The age of felled tree can be determined if the stump shows annual rings.
- The C.A.I and M.A.I do not coincide with each other throughout the life of the tree or stand except twice viz., one at the end of first year and the second in the year of the culmination of the M.A,I.
- Value of C.A.I may be zero or negative but the value of M.A.I is never zero. The two subsequent value of M.A.I are equal.
- Formulae to determine the **diameter and volume increment** percent are given below:

For diameter increment percent	For volume increment percent
Compound interest formula, $P = 100\left[\left(\frac{D}{d}\right)^{1/n} - 1\right]$	Compound interest formula, $P = 100\left[\left(\frac{V}{v}\right)^{1/n} - 1\right]$
Pressler's formula, $p = \frac{200\,(D-d)}{n\,(D+d)}$	Pressler's formula, $p = \frac{200\,(D-d)}{n} \times \frac{(V-v)}{n\,(V+v)}$

- **Schneider's formula** for increment percent is given below:

$$p = \frac{400}{\pi D}$$

Where, D is present diameter & n is the number of rings in outermost cm in radius.

- **Stump analysis** is the analysis of a stump cross-section by measuring annual rings in order to estimate the age of the tree and its past rate of diameter and basal area growth.

- **Stem analysis** is defined as the analysis of a complete stem by measuring annual rings on a number of cross-section at different heights in order to determine its past rates growth.
- **Increment boring** is defined as the boring of a tree stem with Pressler's increment or any other borer to determine increment of trees with annual rings.
- Diameter corresponding to the mean basal area of a uniform, generally pure crop is called **crop diameter.**
- Diameter corresponding to the mean basal area of a group of trees or a stand is called **mean diameter**. Mean basal area is calculated by using the given formula:

  $$\text{Mean basel area} = \frac{n_1 s_1 + n_2 s_2 + n_3 s_3 + ...}{n_1 + n_2 + n_3 + ...}$$

  Where, n_1, n_2, n_3 are no. of trees in each dia. Class. & s_1, s_2, s_3 - basal area of mean trees.
- Mean diameter could also be calculated as an arithmetic average of the diameter of trees. This is given below:

  $$\text{Mean diameter } D = \frac{n_1 d_1 + n_2 d_2 + n_3 d_3 + ...}{n_1 + n_2 + n_3 + ...}$$

  Where d_1, d_2, d_3 are number of trees in each diameter class.
- Diameter corresponding to the mean basal area of the 250 biggest trees per hectare in a uniform, generally pure crop is called top diameter.
- **Crop height** is the average height of a regular crop as determined by Lorey's formula given below:

  $$\text{Crop height} = \frac{h_1 s_1 + h_2 s_2 + h_3 s_3 + ...}{s_1 + s_2 + s_3 + ...}$$

  Where h_1, h_2, h_3 are mean height of trees of these diameter class.
- Height corresponding to the mean diameter of a group of trees or the crop diameter of a stand is called **mean height**.
- Height corresponding to the mean diameter of the 250 biggest trees per hectare in a uniform, generally pure crop is called **top height**.
- **Crop age** is defined as the age of a regular crop corresponding to its crop diameter. The following formula is used for determined crop age for even-aged crop:

  $$\text{Crop age} = \frac{a_1 s_1 + a_2 s_2 + a_3 s_3 + ...}{s_1 + s_2 + s_3 + ...}$$

Where, a_1, a_2, a_3 are age of the even-aged groups. & s_1, s_2, s_3 are basal area of the even-aged.

- **Schlich** defined mean age of uneven-aged wood as "period which an even-aged wood requires to produce same volume as the uneven-aged wood."
- According to Indian forest, mean age is defined as the average age of dominant trees in a crop.
- Table showing various formulae for calculation of mean age of uneven-aged crop:

Formula	Given by
Mean age $= \dfrac{V_1 + V_2 + V_3 + ...}{V_{1/a_1} + V_{2/a_1} + V_{3/a_1} + ...}$	Smalian & Heyer
Mean age $= \dfrac{a_1 s_1 + a_2 s_2 + a_3 s_3}{S}$	-
Mean age $= \dfrac{a_1 n_1 + a_2 n_2 + a_3 n_3 + ..}{n_1 + n_2 + n_3 + ...}$	Andre
Mean age $= \dfrac{a_1 m_1 + a_2 m_2 + a_3 m_3 + ...}{m_1 + m_2 + m_3 + ...}$	Gumpel
Mean age $= \dfrac{a_1 + a_2 + a_3 + ...}{n}$	-

- **Sample plot** is a plot chosen as a representative of a larger area. In India mostly rectangular plot are laid and size of plot 0.1 to 0.2 hectare is considered sufficient.
- About 600 plots for one species are considered enough for preparation of yield tables.
- **The surround** is defined as an area maintained round an experimental or sample plot to ensure that the latter is not being affected by the treatment applied to the area outside the plot. It is also known as **isolation or buffer strip**. All trees in the surroundings are generally marked by white paint rings.
- In India, sample plots are generally thinned to C or D grade ordinary thinning.

- Table showing various formulae for calculation sample plot volume:

Methods	Formula	Based on	Drawbacks
Hossfeld	$V = n_1 v_1 + n_2 v_2 + n_3 v_3$	1 inch diameter class group	Requires fellings large number of tree
Huber	-	Equal number of dia. class	Varying no. of trees in each group
Urich	$V = \frac{Vs \times S_1}{S_2}$ S_1 stand basal area S_2- total basal area of sample tree V_s - total volume of sample tree	Same no. of tree select from each group	Artificial grouping
Hartig	$V = \frac{V \times S}{s}$ v - volume of sample tree S - group basal area s - basal area of sample tree	Each group have equal volume. This method is used in temporary sample plots	More computation work
Block	-	Group have highest to lowest diameter. This method is used by FRI, Dehra Dun.	-
Draut	$V = C \times v$ v - total volume of all the sample trees C-total no. of trees/ total no. of sample trees	Taking volumes of all sample trees in the group.	-
Form factor method or Prussian institute method	$V = S \times h \times f$ s - basal area of mean tree h - height of mean tree f - form factor	Volume of abstract mean tree calculation	-

- Volume of sample plot can be calculated by using a computer programme named as **FORTRAN**. This programme has been used in FRI for *Eucalyptus hybrid* sample plot volume calculation.

- According to **Loetsch & Haller**, "*Forest inventory*" is the tabulated, reliable and satisfactory tree information, related to the required unit, respectively units, of assessment in hierarchic order. Forest inventory is synonymous with the term **enumeration** used in India.
- **Enumeration** is the counting, singly or together, of individuals of one or more species in a forest crop and their classification by species, size or condition etc.
- There are two main kinds of enumeration viz., Total or Complete and Partial or Sample.
- Total or Complete enumeration means that enumeration of the desired species above the specified diameter limit is carried out over the entire area of the forest unit under consideration.
- Partial or Sample enumeration is of a representative portion of the whole forest.
- The ratio of the sample to the whole population is called **sampling fraction or sampling intensity**. It is expressed as a percentage e.g., 1%, 2%, 5%, 20% etc.
- The choice of the two kinds of enumeration depends upon-
  a. Extent of the covering area
  b. Variation in density and composition
  c. Management intensity & required degree of accuracy
  d. Labour resources, time and available funds.
- **Partial or Sample enumeration** has great advantages over the total enumeration:
  a. Reduced cost and time saving
  b. Relative accuracy
  c. Knowledge of error
  d. Greater scope
- When extensive areas are to be covered partial or sample enumeration is used while where crop composition and density varies greatly (in miscellaneous forest), required greater accuracy, high management intensity, P.B.I and area of shelterwood system then total enumeration is carried out.

- There are two main kinds of sampling used in forest inventories:

Random sampling	Non- random sampling
Simple random sampling	Selective sampling
Stratified random sampling	Systematic sampling
Multi-stage sampling	Sequential sampling
Multiphase sampling	
Sampling with varying probability	
List sampling	

- **Random sampling** is in which sampling units are selected in such a manner that all possible units of the same size have equal chance of being chosen.
- **Simple random** sampling is in which sampling units are selected by some strictly random process from the whole population. It gives good result if the forest crop is uniform.
- The **stratified random sampling** is in which the population is first divided into sub-population of different strata and then units are selected from each of them in proportion to their size. The method of the division of area into homogeneous groups is called stratification.
- **Multi-stage sampling** is in which sampling units are not taken out at one stage but are taken out in two or three stages called multi-stage sampling. In this sampling units gets smaller at each successive stage but following random sampling at each stage of selection.
- **Multiphase sampling** is in which some of the same sampling units are used at different phases of sampling to collect different information or same information by different methods. Two phase sampling is commonly used in forest inventories so it is also called as double sampling. Such as bamboo enumeration & aerial photograph interpretation.
- In the **random sampling** the chance of selection of all sampling units are the same at all times, but in certain population the chance may vary as the sampling proceeds and therefore a modified method of sampling known as sampling with varying probability is used.
- The **Bitterlich or variable plot** or point sampling or probability proportional to size (PPS) and PPP or sampling with probability proportional to prediction are based on sampling with varying probability.

- **List sampling** is another form of sampling with varying probability. This method consists of making a list of sampling units along with their measure of size in any order.
- **Non-random sampling** is the method of sampling in which samples are selected according to the subjective judgement of the observer on the basis of certain guidelines.
- **Selective sampling** is in which samples are selected according to the subjective judgement of the observer. 4 PEA sampling is one of the forms of selective sampling.
- **Systematic sampling** is in which sampling units are selected according to a predetermined pattern. In this method of sampling first unit is selected at random process and other according to a fixed pattern so it is called as systematic sampling with a random start.
- **Sequential sampling** is a method of sampling in which the number of observations in the sample is not predetermined but sampling units are taken successively from a population. Each samples includes all the sampling units of the former sample. This is use to test a hypothesis.
- If the total amount for an inventory is p, then the number of sample unit (n) is calculated by using given formula:

  $$n = \frac{p-c}{r}$$

  Where, c = overhead cost, r = cost per sampling unit
- Sampling units are of two types:
  1. Those having a fixed area
  2. Those having only fixed points.
- The usual shapes of the sampling units in India are plots, strips, topographical units and clusters.
- The term plot is applied to sampling units of small area. Square & rectangular plots are most commonly used in forest survey. But rectangular sample plot are widely used in India having a usual size of the plot 0.1 hectare. Bigger plot of size 1.0 hectare are often laid in mature crops.
- **Strips** having width 20 to 40 m are used in some area to serve as a sample. The width of the strip and the distance between the two strips determine the **intensity of sampling**, which can be calculated by the given formula:

  $$I = \frac{W}{D} \times 100$$

Where, I = sampling intensity, W = strip width, D = central lines distance of two adjacent strips.

- **Topographical sampling** units is defined as a sampling unit whose boundaries are predominantly topographical or natural features such as *nalas*, streams and ridges. Topographical units are mostly used as sampling units in hill forest.
- **Cluster** is defined as a sampling unit which is, in fact a group of small units.
- **Sampling intensity** refers to the percentage of the area of the population to be included in the sample. In order to determine optimum intensity in various forests the sampling error should be less than **10%**.

Forest type	Sampling percentage
Tropical wet evergreen	10
Tropical moist deciduous	2.5
Sub-tropical pine forest	5

Terrain	Method	Sampling percentage
Plain	Strip sampling	5 to 10
	Line plot survey	2 to 5
Hills	Topographical units	20 to 25 (forest area >2000 ha)

- In the plains the number of plots is fixed on the given following criteria:

Net area of forest unit	Enumeration intensity
Upto 10 ha	Total Enumeration
11-50 ha	Min. 30 circular plots
51 ha or more	30-100 plots.

- **National Inventory Design** concept was given by Forest Survey of India.
- There are two main sources of errors in sampling viz., sampling error and non-sampling error.
- **Accuracy of an inventory** refers to the size of total error of the sampling in inventory and includes error due to bias. While precision

refers only to the size of the sampling error and excludes the error due to bias.

- **Sampling error** refers to the difference between the population mean (*M*) and sample mean (*Y*). It is denoted by *E*.

$$E = M - Y$$

Enumeration types	Number of enumerator	Number of *mazdoors*
Total enumeration	1	7
Liner strip enumeration	1	13-14
Partial enumeration (Hills)	1	7

- **Basal area** is expressed in m^2 while land area is expressed in hectare (10,000 m^2).
- **Point sampling** refers that counting from a random point the number of trees whose breast height cross-section exceeds a certain critical angle which is multiplied by a constant factor to give an unbiased estimate of basal area per hectare. This is proved by **Bitterlich.** This technic is called by several names such as *angle count cruising, pointless cruising, point sampling, variable plot cruising, P.P.S (probability proportional to size) sampling* and *polyareal plot sampling.*
- There are two main kinds of point sampling:
  1. Horizontal point sampling – widely used in India.
  2. Vertical point sampling.
- The process of sampling in which a series of sampling points are selected randomly or systematically distributed over the whole area, when trees around these points are viewed through any angle-gauge at breast height then the tree having an angle greater than critical angle of angle-gauge are counted. This process is called **horizontal point sampling.**
- The number of trees counted and multiplied by a constant factor (depends only on angle size) gives the basal area per hectare.

$$2\sin\frac{\theta}{2} = \frac{D_a}{R_a} = \frac{D_b}{R_b} \text{ and } k = \frac{D}{R} = 2\sin\frac{\theta}{2}$$

- In other words the ratio between the tree basal area & area of circular plot is $k^2/4$.

- The multiplying factor associated with any point sampling instrument is called as basal area factor (BAF) or simply is called **area factor**. It is independent of tree diameter and denoted by F ($F = 2500k^2$).
- The factor which indicates the distance at which a given angle gauge will exactly cover a one metre wide rectangular target called **calibration distance factor**. It is 100 times of plot radius factor. It is denoted by C.
- Tree factor is the number of trees per hectare (or acre) by each tree tallied. It is denoted by F_t. where leave line $\boldsymbol{F_t = BAF/B_i}$ (B_i is basal area of a particular tree).
- **Horizontal point sampling** is used for determination of the following-
  a. Basal area per hectare (BA) = $\boldsymbol{BAF \times n}$ where, n-tally trees
  b. Numberofstemperhectare(N) = $\dfrac{\text{Basal Area Factor of prims}}{\text{Total Basal area of the tally trees}}$
  c. Volume per hectare (V) = **basal area × stand form height.**
- The following instruments are used in horizontal point sampling:
  1. Simple angle gauge
  2. Wedge prism
  3. Spiegel Relaskop
  4. Wide-scale Relaskop- for measuring large trees.
  5. Tele Relaskop
- **Wedge prism:** It is a wedge-shaped piece of glass and works on bending of light rays passing through the prism. This property is helped in counting of tally trees. The tree is counted if the image overlaps the directly viewed tree. The tree is not counted if there is a gap. The Basal Area Factor of wedge-prism is calculated on the basis of the critical angle of prism.
- The formula of basal area factor of wedge-prism is given below:

$$\text{BAF} = \frac{10{,}000}{1 + 4\left(\dfrac{L}{D}\right)^2}$$

- No slope correction is required upto an angle of **18°**. A slope angle of **25°** cause an error of about **10%** in counting of tally trees by wedge prism.

- It is necessary that the prism is held in a **vertical** position in **level terrain** and at right angle to the line of sight on **sloping ground**.
- The overlapping image is called a fully tally tree and the marginal touching image is called **borderline or half tally**. The fully tally trees are counted as one while two half tally trees makes one full tally tree.
- When total tally trees is multiplied by the basal area factor (BAF) gives total basal area (m^2) per hectare. And when this result is multiplied by the area of the plot then it gives basal area inside the plot.
- **Spiegel Relaskop** is widely used in point sampling. It has following scales:
  - 20 meter height band
  - 25 meter height band
  - 30 meter height band
  - Band 1
  - Band 2
  - Four narrow band [2 white + 2 black]
  - Range finder bands
- Band 1 plus 4 narrow bands are collectively called **band 4**.
- Spiegel Relaskop is used to find the following parameters:
  - Range finding
  - Height measurement–height is obtained by multiplying distance to the tree by slope.
  - Slope measurement–height bands gives the angle of slope.
  - Diameter measurement- bands 1 & 4 narrow bands are used for diameter measurement.
  - Determined basal area per hectare- total tally trees × BAF
  - Number of stem per hectare- **total tally trees × BAF/g** where, g-basal area of average tree of the diameter class. The result is the number of stems of that class per hectare. The sum of the number of the different diameter classes gives total number of stem per hectare.
  - Volume of individual tree-
  - Form-factor & form-height- Pressler has given following formula for volume = $\mathbf{2/3 \times} \boldsymbol{gh_1}$.
  - Volume of stands- $\boldsymbol{V = G \times FH}$, where G-basal area per ha. FH-weighted avg. form height.

- **Tele Relaskop** is used to calculate accurate height & diameter of single standing tree.
- There are several methods to determine the volume of ***standing*** trees:
  1. Hohenadl's formula
  2. The 'odd ten per cent's method
  3. The 'Any visible diameter's method
  4. The rope method
- **Non-sampling errors** in horizontal point sampling are given below:
  a. Instrumental errors and personal bias
  b. Deviation from the prescribed sample point
  c. Bias due to concealed trees
  d. Error due to incorrect inclusion or exclusion of borderline trees
  e. Bias due to imaginary circle extending outside the stand.
- A process of sampling in which all trees appearing more than a critical angle are counted with in a full 360° sweep around sample point is called **vertical point sampling**. Here, the value of critical angle is 45° fixed for tally tree.
- **Conimeter** is based on the principle of vertical point sampling.
- **1 diopter** is equal to an angle of **0.57**° or 34.36 minutes.

CHAPTER 8

Remote Sensing

- Term remote sensing was coined by **Ms. Evelyn Pruitt** in 1960s.
- Father of Indian remote sensing is **Pisharoth Rama Pisharoty.**
- First aerial photograph was taken in 1858 by **Felix Tournachon**, known as Nadar, from a tethered balloon over the Bievre Valley in France. Those photographs no longer exist. The oldest surviving aerial photograph is the one of Boston, taken by **James Wallace Black** in 1860, using a tethered balloon.
- India launched its first satellite **"ARYABHATA"** in 1975 with the help of **"KAPUSTIN YAR"** Russian rocket launch site.
- The **Indian Space Research Organization (ISRO)** is the space research and exploration agency of government of India (GOI), it was formed in 1969 and headquartered in Bengaluru (Karnataka).
- In 1957, Russia became the first country to launch an artificial satellite **"SPUTNIK 1"** rocketed by **"SPUTNIK 8K71PS"**.
- **Vikram Sarabhai** is the father of Indian space programme.
- According to **Lillesand (1970)**, "Remote sensing is the art and science of obtaining the information about an object, area or phenomena through the analysis of data acquired by a device that is not in direct contact with the object, area or phenomena under investigation."
- The most common and important sources of energy used in remote sensing is the sun.
- The electromagnetic radiation occuring as a continuum of wavelength (λ) and frequency from short wavelength and high frequency cosmic waves to long wavelength and low frequency radio waves, is called **electromagnetic spectrum (EMS)**.
- The bands of spectrum in which there is little or no absorption found are called **atmospheric windows.** It exists in wavelength (λ) 0.4-1.5 μm, 3.0-5.5 μm, 8.0-14 μm, and wavelength larger than 100 μm.

- Remote sensing is classified into two categories viz., aerial remote sensing and space remote sensing on the basis of distance of platform from the earth.
- **Aerial remote sensing**- Aerial remote sensing is also called as aerial photography. Aerial photography is the science of obtaining the photographs from the air using various platforms like as balloons, aircrafts etc. for getting information about the surface of the earth.
- **Space remote sensing**- Space remote sensing is a method of remote sensing in which cameras, sensors or other devices are attached to the satellite orbiting round the earth, take the digital photographs of earth & resources and after converting these digital photographs into photographs by computer, required information is obtained.
- In aerial remote sensing the platform height varies from **5000 m** to **15000 m**. While in space remote sensing it varies from **300 km** to **1000 km**.

Aerial photographs show	Maps shows
Perspective projection of earth surface	Orthographical projection of earth surface
Actual image of the object & the ground features.	Objects and ground features show by conventional symbol.
3-D view of earth	Only two dimensional (2-D) view.

- Classification of aerial photographs are given below:

On the basis of film used			
Photographs name	**Film used**	**Remarks**	**Used for**
Panchromatic Black & White photograph	Panchromatic film	This film has low sensitivity for green region	Not suitable for plant species identification
Infra-red (IR) black & white photograph	Infra-red (IR) film with proper filter	This film is used with red filter for identification.	Identification of broad leaved & coniferous tree
Colour photograph	Colour films	Colour films has three emulsion layers (blue, green and red). green produces burnt magenta while red produced cyan and blue produces yellow colour.	Identification of species and diseases.

Infra-red colour photograph	Kodak Aero-chrome Infra-red	Also called false colour photograph because object shows different colour from their actual colour	Green vegetation is seen red to magenta colour in this film.

On the basis of the device used	
Photographs name	**Device used**
Single lens photograph	Single lens camera
Multi-spectral photograph	Single camera with more than one lens
Multi-spectral imagery	Optical-mechanical scanner

On the basis of scale of photograph		
Photographs name	**Scale**	**Used for**
Very large scale photograph	1: 5,000 to 1: 10,000	Working plan purposes, detailed management inventory, exploitation planning, insect & diseases survey and small height classes classification.
large scale photograph	1: 10,000 to 1: 20,000	
Medium scale photograph	1: 20,000 to 1: 40,000	Forest inventory preparation, detailed land use classification, forest stratifications, broad height classification.
Small scale photograph	1: 40,000 to 1: 70,0000	Broad land use classification, forest type identification.
Very small scale photograph	1: 70,0000	

On the basis of scale of position of optical axis of camera	
Photograph name	**Deviation**
Vertical photograph	Upto 4° from perpendicular position
Oblique photograph	Low oblique- upto 30°
	High oblique- more than 30°

- Silent features of vertical & oblique photograph is given below:

Vertical photograph	Oblique photograph
Can be used as map Area covered in rectangular form. Can be taken in any type of country	Can not be used as map Area covered in trapezoidal form. Only taken in terrain.

- In India, following given agencies are responsible for aerial photography:
  1. Indian Air Force (IAF)- New Delhi
  2. Messers Air Survey Company of India (Pvt.) Limited, Dumdum-Kolkata
  3. National Remote Sensing Agency (NRSA), Hyderabad
- The photography of **sensitive or restricted area** is carried out by Indian Air Force (IAF).
- The Surveyor General's office for aerial photography is situated in **Dehra Dun.**
- Aerial photography lens are divided into 4 categories on their area coverage:

Lens name	Coverage angle
Narrow angle	50°
Normal angle	60°
Wide angle	90°
Super wide angle	120°

- A **yellow filter** absorbs blue energy incident on it and transmits green & red energy. While **red filter** absorbs blue, violet and green energy and transmits higher wavelength of visible spectrum.
- Most commonly used filter in forestry is **yellow filter** while for water bodies **red filter** is used. Panchromatic film are exposed through minus blue or red filters to reduce the effects of atmospheric haze.
- There is about 30% overlap in between the photographs taken from two consecutive flight lines. While drift and crab should not exceed 10% of the width of the photograph.
- The ideal time for taking a photograph is **9 to 11 A.M** and **1 to 2 P.M.**

Vegetaional form	Optimum season for aerial photography
Coniferous forests in Himalayan	October–November
Mixed deciduous forests (no teak & sal)	December–February
Sal forest	Mid-March–Mid April
Teak forest	September–December
Tropical evergreen & moist deciduous	January–February

CHAPTER 9

Plant Physiology

- The study of various activities and metabolism of plants is known as **plant physiology**.
- **Stephan Hales** is known as father of plant physiology.
- **J.C. Bose** is known as father of Indian plant physiology.
- Water forms **80-90%** of fresh weight of plant body.
- The movement of molecules or atoms or ions of a materials from an area of higher concentration to an area of their lower concentration is called **diffusion**.
- The diffused molecules or ions exert a pressure on the substance or the substance or medium in which diffusion takes place, known as **diffusion pressure**.
- Diffusion rate ⟶ **Gas > Liquid > Solid.**
- Diffusion pressure of a pure solvent (**1236 atm**) is always higher than its solution.
- Exchanges of gases like CO_2, O_2 and distribution of hormones in plants takes place through diffusion.
- The process of **osmosis** is a special type of diffusion. The process of transpiration is a diffusion process.
- Osmosis is defined as the special diffusion of solvent from the solution of lower concentration to the solution of higher concentration through the **semi-permeable** membrane.
- Osmosis was discovered by **Abbe Nollet.** The detailed explanation of osmosis was given by **Traube** and **Duterochat**.
- The water moving into the cell during the osmosis is called **endosmosis**.
- The water moving out the cell during the osmosis is called **exosmosis**.
- The exchange of material in and out through the membrane is called as **permeability**.

- Membrane which is permeable for both solute and solvent is called **permeable membrane**.
- Membrane which is permeable for solvent and do not allow the passage of solute is called **Semi-permeable membrane**. Such as Cellophane and Copperferrocyanide membrane, goat bladder, parchment paper etc.
- Membrane which allow some selective solutes to pass through them along with the solvent molecules is called **selective permeable membrane**. Such as cell membrane, tonoplast and organellar membrane. These membranes are permeable for gases and impermeable for polysaccharides and proteins.
- The pressure developed in a solution when solution and water are separated by semi-permeable membrane is called **osmotic pressure**.
- The osmotic pressure of **pure water is zero**. According to **Hariss** the osmotic pressure is highest in leaves and lowest in roots.
- The highest osmotic pressure is found in halophytes while lowest in hydrophytes.
- Order of osmotic pressure in plants, Hydrophytes < mesophytes < xerophytes < halophytes.
- Osmotic pressure of a solution is measured by osmometer. First osmometer was made by **Pfeffer**.
- The formula of Vont Hoff for measuring osmotic pressure is **OP = mRT**.
- The osmotic pressure of electrolytes is higher than that of non-electrolytes.
- Water moves from lower osmotic pressure towards the higher osmotic pressure.
- Root hairs of the roots absorb water from the soil through the process of osmosis.
- The **opening and closing of stomata** is also dependent on the process of osmosis.
- The leaves of *Mimosa pudica* ("Touch me Not") are drooping down due to the change in turgor pressure regulated by osmosis.
- When a cell is immersed in water, then water enter into the cell because osmotic pressure of the cell sap is higher. The cell content press upon the wall or develop a pressure against the cell wall which is called **turgor pressure**.

- A flaccid cell has zero turgor pressure while a fully turgid cell has highest turgor pressure.
- Turgor pressure regulate plant movements and maintain the normal shape of the cell.
- The difference between the diffusion pressure of the solution and its pure solvent at particular temperature is called diffusion pressure deficit (DPD). Diffusion pressure deficit also is called as **suction pressure.**
- **DPD = OP – TP or WP.** Where OP means osmotic pressure and WP means wall pressure.
- Diffusion pressure deficit (DPD) for fully turgid cell is zero. While in case of flaccid cell, DPD is equal to osmotic pressure (DPD = OP).
- DPD for plasmolysed cell is equal to OP + TP. So that DPD of the plasmolysed cell is greater than osmotic pressure.
- The difference between the free energy of molecules of water and free energy of the solution is called as **water potential**. It is denoted by Ψ_w. Water always moves from higher water potential to lower water potential.
- Water potential of pure water is maximum. Water potential is equal to DPD but opposite in sign (**Ψ_w = –DPD**). Its value is negative.

$$\Psi_w = \Psi_s + \Psi_p$$

  Where Ψ_s = osmotic potential, Ψ_p = pressure potential and Ψ_w = water potential.

- Ψ_s has positive sign (+ve) while Ψ_w and Ψ_p have negative sign (–ve).
- Adsorption of undissolved liquid by any solid material is called **imbibition**.

  Imbibition power ⟶ Agar-Agar > Pectin > Protein > Starch > Cellulose

- Breaking of seed coat during the seed germination is due to imbibition process.
- Newly formed wood swells up in rainy season due to imbibition process.
- During the seed germination absorption of water takes place through imbibition.
- **Resurrection** in many plants like as Selaginella, lichens and mosses takes place due to the process of imbibition.

Movement of water molecules		
Higher D.P	**⟶**	**lower D.P**
Lower O.P	**⟶**	**higher O.P**
Lower DPD	**⟶**	**higher DPD**
Higher (less –ve) Ψ_w	**⟶**	**lower (more –ve) Ψ_w**
Higher T.P	**⟶**	**lower T.P**
Hypotonic	**⟶**	**hypertonic**
Lower conc. of solution	**⟶**	**higher conc. of solution.**

- Water potential is measured in **bars or Pascal**. Value of water potential in some special case are shown in the table:

Types of cell	Value of Ψ_w
Full turgid cell	Zero
Flaccid cell	Negative
Plasmolysed cell	Negative

- Forms of water which reaches at the soil water table due to the gravitational force after the rainfall and not available to plant is called **gravitational water**. Some plants can absorbed this water such as *Calotropis, Prosopis and Capparis* etc.
- Thin film of water which is tightly held by the soil particles is called **hygroscopic water**. This water is not available to the plant.
- The amount of water present in the chemical compounds, which are present in the particles of soil is called **chemically combined water**. This water is not available to the plant.
- Water exits between the soil particles in small capillary pores is called **capillary water**. It is the most available form to the plants. Plants only absorbe this form of water.
- Water vapour present in the air, which can be absorbed by hanging roots of the epiphytes due to velamen tissue is called **atmospheric humidity**.
- Total amount of water present in the soil is called as **holard**.

  Holard = Chresard + Echard
- **Chresard**- this is the water available to the plants.
- **Echard**- this water is not available to the plants.

- Maximum absorption of water takes place from root hair region. While maximum absorption of mineral takes place from meristematic region.
- The movement of water from cell to cell through plasmodesmata is called **symplast or living path**.
- There are two way of water absorption in plants viz. active water absorption and passive water absorption.
- Osmotic active method of water absorption was given by **Atkins & Priestley**. While non-osmotic active method of water absorption was given by **Thieman, Bennet-Clark.**
- About **90%** of water in plants is absorbed by passive method.
- Term active and passive absorption was proposed by **Renner**.
- Upward movement of water against the gravitational force upto top parts of the plants is called **ascent of sap**.
- Girdling or ringing experiment was given by Malpighi, Hartig and Stephen hales.

Theory	Explained	Given by
Relay pump theory	Ascent of sap	Godlewski
Pulsation theory	Ascent of sap	J.C. Bose
Root pressure theory	Ascent of sap	Priestley
Transpiration pull & cohesion force theory	Ascent of sap	Dixon & Jolly
Capillary force theory	Ascent of sap	Boehm
Chain theory	Ascent of sap	Jamin
Imbibition force theory	Ascent of sap	Unger & Von Sachs
Cytochrome pump theory	Mineral absorption	Lundegardh & Burstorm
Carrier concept theory	Mineral absorption	Vanden Honert
Protein- Lecithin theory	Mineral absorption	Bennet & Clark
Lock & key theory	Enzyme action	Emil Fischer
Induced fit theory	Enzyme action	D.E Koshland
Chemiosmotic theory	Photophosphorylation	Peter Mitchell

- **Transpiration pull & cohesion force theory** is most accepted or universally accepted theory of ascent of sap.

- Translocation of food mainly occurs in the form of **sucrose** (non-reducing sugar).
- Generally green photosynthetic plant parts, which forms produces food act as **source** like leaves while non-photosynthetic parts unable to form their food like roots, shoots and fruits act as **sink.**
- **Pressure flow** or mass flow hypothesis of food translocation was given by **E. Munch** in 1930. This is most accepted theory of food conduction in plants.
- Loss of water in the form of vapour, from the aerial parts of living plants is known as **transpiration**.
- According to **Curtis, transpiration is an essential evil.**
- According to **Steward, transpiration is an unavoidable evil.**
- The minimum transpiration is found in succulent xerophytes & maximum in mesophytes while absent in submerged hydrophytes.

Transpiration type	Takes place through	Transpiration percent
Stomatal	Stomata	80-90%
Foliar	Leaves	–
Cuticular	Cuticle	9%
Lenticular	Lenticels	0.1-1%

- **Stomata** are found on the aerial delicate organs and outer surface of the leaves. **Stomatal pore** is surrounded by two specialised epidermal cells called as guard cell (kidney shaped).
- The structure of guard cell in monocots (Gramineae) is **dumbled shaped**.
- The outer wall of the **guard cell** is thin & elastic, while inner wall is thick and non-elastic.
- **Sunken stomata** are found in xerophytic plants.
- Active potassium (K^+) and hydrogen (H^+) theory was given by **Levitt**. This is the modern and most acceptable theory for stomatal movement.
- Plant hormone **Abscisic acid** (ABA) help in closing of stomata.
- Chemical substances which reduce the rate of transpiration are known as **antitranspirants**. Such as Phenyl Mercuric Acetate (PMA), Aspirin, Abscisic acid (ABA), CO_2, Oil and low viscous wax.
- Number of stomata present in unit area is called as **stomatal index**.

- Loss of water from the uninjured part or margin of leaves of the plant in the form of water droplets is called as **guttation**. Term guttation was coined by **Burgerstein**.
- Guttation takes place through a special pore like structure called **hydathodes or water stomata**. Process of guttation is governed by root pressure.
- Fast flow of liquid from the injured or cut parts of the plants is called as **bleeding or exudation**. Process of exudation is governed by root pressure. Sugar is obtained from the sugar maple tree by exudation. The highest bleeding is found in *Caryota urens* (Toddy palm-about 50 liters per day).
- Dropping of the soft parts of the pants due to loss of turgidity in their cells is called **wilting**.
- About 50-60 elements are present in plant body but only **17** elements are considered as **essential elements**.
- Criteria of essentiality of minerals was given by **Arnon** in 1935. According to Arnon criteria of essentiality of minerals :
    - The element must be necessary for normal growth and reproduction of all plants.
    - The requirement of element must be specific for plant life. That is indespensible element to plant.
    - The element must be directly involved in metabolism of plants.

Element	Required concentration	Example
Major element or Macro nutrients	1-10 μg L^{-1}/ 10 m mole kg^{-1}	**C,H, O, N, K, S, Ca, Mg, P**
Minor element or Micro nutrients	<1-10 μg L^{-1}/10 m mole kg^{-1}	**Fe, Cu, Zn, B, Cl, Mn, Mo**

- Mobile elements can be remobilized from senescent (older) organ to young plant parts and deficiency symptoms first appear in older plant parts. Example of mobile elements are **N, P, K** and **Mg.**
- Immobile elements can not be remobilized from senescent (older) organ to young plant parts and deficiency symptoms first appear in younger plant parts. Example of immobile elements are **B, S, Ca, Co** and **Fe**.

Element	First deficiency symptom shown by	Examples
Mobile element	Older plant parts	N, P, K, Mg
Immobile element	Younger plant parts	B, S, Ca. Co

- Table showing mineral nutrition in plants:

Name	Absorption form	Role/function	Deficiency symptoms
Nitrogen (N)	NO^{3-} (Nitrate)	1. Constituent of plant hormones (IAA) and ATPs. 2.Present in porphyrins of chlorophylls & cytochromes.	1. Chlorosis in older leaves 2. Anthocynin formed in stem 3. Late flowering & plant becomes more susceptible to fungal diseases due to excess of Nitrogen
Sulphur (S)	SO^{4-} (Sulphate)	1. Parts of cystinine, cysteine & methionine amino acids. 2. Vit. biotene,thiamine, Co-A in respiration 3. Role in oil synthesis, chlorophyll synthesis & part of ferredoxin. 4. Root nodule formation	1. Chlorosis or yellowing in younger leaves with anthocyanin. 2. Tips & margins of leaves curves inwardly "tea yellow disease"
Phosphorus (P)	$H_2PO_4^-$ (Orthophosphate)	1. Very important in RNA, DNA, Phospholipid (cell membrane) NADP (Co-enzyme) ATP (Energy reaction). 2. Very important in photosynthesis (NADP), Protein synthesis.	1.Premature leaf fall, necrosis and anthocyanin formation.
Calcium (Ca)	Ca^{2+}	1. Mechanical strength because Ca is constituent of middle lamella. 2. Permeability of bio membrane is maintained. 3. Stability of chromosome structure & in spindle formation.	1. Disintegration of growing apices 2. Irregular cell divisions and death of meristem.
Molybdenum (Mo)	MoO_4^{2-} (molybdate)	1. Role as prosthetic group of nitrate reductase and nitrogenase in nitrogen metabolism.	1. Whip tail of cauliflower
Potassium (K)	K^+ (monovalent cation)	1. Key role in stomatal movement & transpiration.	1. Mottled chlorosis 2. Die-back disease 3. Decrease the apical dominance
Magnesium (Mg)	Mg^{2+}	1. Constituent of chlorophyll. 2. Activator of many enzymes in carbohydrate metabolism.	1. Interveinal chlorosis on large scale and formation of anthocyanin in older leaves.

Iron (Fe)	Fe^{2+} (Ferrus)	1. Absorption from acidic soil 2. Iron porphyrin protein for cytochromes, Peroxidase & catalases 3. Impotrant for ferredoxin, biological N_2 fixation & ETS 4. Essential for chlorophyll synthesis	1. Rapid interveinal chlorosis.
Manganese (Mn)	Mn^{2+}	1. Essential for O_2 evolution and photolysis of water in light reaction.	1. Marsh spot of pea, and grey speak of oat. 2. Chlorosis in young & older leaves.
Boron (B)	$H_3BO_3^-$ (Borate)	1. Boron is only micronutrient, which is not associated with enzymes. 2. Required for uptake and utilization of Ca^{+2}, membrane functioning, cell elongation and differentiation and carbohydrate translocation. 3. Essential in pollen tube formation. 4. Lethal effect at carbohydrate metabolic site.	1. Stem and root tips (apex) dies. 2. Top rotten in tobacco, water core in turnip, brown heart rot of beets, heart rot in carrot & marigold, fibres in apple fruit.
Copper (Cu)	Cu^{2+}	1. Vit.-C (Ascorbic Acid) formation.	1. Die-back of citrus and other fruit trees exanthema in the trees. 2. Reclamation disease of cereals & legume crops.
Zinc (Zn)	Zn^{2+}	1. Specific role in Auxin (IAA) synthesis. 2. Activator of Carbonic anhydrase, carboxylases, alcohol dehydrogenase and peptidase.	1. Mottle leaf disease in fruit trees 'little leaf disease'. 2. Khaira disease of paddy' Rosset disease in walnut. 3. White bud disease in maize.

- C, H, O, N and P are main constituents of protoplasm, so called as **protoplasmic elements**.
- Hydroponics/solution culture/soil less growth/tank farming and ash analysis are the techniques which determinates the role of nutrients in plants. This concept was given by **Geriack**.
- Plants grown in moistened air with nutrients is called **aeroponics**.

Plant hormone	Discover by	Bio-assay	Effects
Auxin basipetal & polar	F.W. Went (1928)	1. Tryptophan amino acid 2. Avena curvature test 3. Root growth inhibition test	1. Apical dominance 2. Callus formation 3. Parthenocarpy 4. Femaleness in plants
Gibberellins (GA_3) $C_{19}H_{22}O_6$	Yabuta & Sumiki (1935)	1. Mevalonic acid pathway 2. Dwarf pea & maize test 3. Alpha amylase activity test	1. Internodes elongation 2. Elongation of genetic dwarf plants 3. Flowering in LDP, in short light duration 3. Vernalization 4. Breaking of dormancy 5. Seed germination
Cytokinins basipetal & polar	Miller	1. Tobacco pith cell division test 2. Chlorophyll preservation test 3. Soyabean and Radish cotyledon cell division test	1. Promote cell division 2. Cambium formation 3. Delay in senescence 4. Flowering in SDP
Abscisic Acid stress hormone $C_{15}H_{20}O_4$	Addicott & Okhuma (1963)	Mevalonic acid pathway	1. Induce abscission 2. Induce bud & seed dormancy 3. Stomatal closing
Ethylene C_2H_4	Burg	Methionine	1. Fruit ripening hormone 2. Triple response on stem 3. Inhibit the polar movement of auxin

- **Chlormequat (CCC or cycocel)** is a growth inhibitor which is used in bonsai.
- Term hormone was given by **Starling** while phytohormone by **Thieman.**
- 2, 4-D, and 2, 4, 5-T are called **agent orange** and used as selective weed killer. **TIBA** (Tri-Ido Benzoic Acid) is an antiauxin.
- High auxin and low cytokinin favours **root formation** while High Cytokinin and low auxin favours **shoot differentiation.**
- The first natural cytokinin was identified & crystalized from immature corn grains by **Letham** and named as **zeatin.**
- **Indole butyric acid (IBA)** is root initiating hormone in cutting. **Indole acetic acid (IAA)** is a natural auxin. NAA, 2, 4-D, 2, 4, 5-T, IPA, IBA and Maleic hydrazide are synthetic auxins.

- The relative length of day and night is called as **photoperiod**. The response of plants to the photoperiod, expressed in the form of flowering is called as **photoperiodism**. Effect or requirement of relative length of day & night on the flowering of plant is called as **photoperiodism.**
- Photoperiodism was first discovered by **Garner & Allard** on tobacco (Maryland mammoth).
- Garner & Allard classified the plants into **Short day plant (SDP), long day plant (LDP) and Day neutral plant (DNP).**

Plants	Required photoperiod	Examples
Short day plant	1. Shorter than their critical day length 2. Dark period must be continuous 3. Called as long night plant 4. Dark period is critical	*Tobacco, Soyabean, Viola, Xanthium, Chrysanthemum, Cannabis, Coleus, Chenopodium, Mustard, Dahlia, Sugarcane, Strawberry, Cosmos And Rice.*
Long day plant	1. Longer than their critical day length 2. Light period is critical	Henbane, Spinach, Avena, Wheat, Sugarbeets, Raddish, Larkspur, Barley and Potato
Day neutral plant	Not specific	Zea, Cotton, Tomato, Sunflower and Cucumber

- ***Florigen,*** a hypothetical hormone was discovered by **Chailakhyan**.
- Phytochrome pigment responsible for flowering is discovered by **Borthwick & Hendricks** and term phytochrome was given by **Butler**.

Phytochrome	Absorption range	Induce flowering in
Phytochrome red (Pr)	630-670 nm	Short day plant (SDP)
Phytochrome far red (Pfr)	720-740 nm	Long day plant (LDP)

- Effect of low temperature on the initiation and development of flower is called **vernalization**. This phenomena was discovered by **Lysenko**.
- **Enzymes** are biocatalysts made up of proteins (except ribozyme), which increases the rate of biochemical reactions by lowering down the activation energy.
- Term enzyme was given by **Kuhne** and first isolated enzyme is zymase (from yeast) discovered by **Buchner**.
- First discover ribozyme was L_{19} RNAase by **T. Cech**. The first purified and crystalized enzyme was urease **(by J.B Sumner)**.

- Protein part of conjugated enzyme is called **apoenzyme**. Conjugated enzyme (made up of protein + non-protein) is also known as **holoenzyme**.
- **Km** constant of an enzyme, is the concentration of substrate at which rate of reaction of that enzyme attains half of its maximum velocity. It is given by **Michaelis & Menten**. The value of Km should be lower for an enzyme.
- **ATP** was discovered by Lohman, while importance of ATP in metabolism by Lipman

Substrate	Equal to
1 gram fat	9.8 K.Cal.
1 gram protein	4.8 K.Cal.
1 gram carbohydrate	4.4 K.Cal.

- **J. Priestly** carried out experiment on Bell jar. Law of limiting factor was given by **F.F. Blackman** in 1905.
- **Van Niel** proved that oxygen released from water and oxygen of glucose comes from carbon di oxide (CO_2).
- **Ruben, Hassid & Kamen** used **O^{18}** to experimentally show that O_2 in photosynthesis released from water.
- **Moll** proved that CO_2 is essential for photosynthesis by half leaf experiment.
- **Photosynthesis** is a photo-biochemical process (anabolic & catabolic) in which organic compounds (carbohydrates) are synthesised from the inorganic raw material (H_2O & CO_2) in the presence of light & pigments O_2 is evolved as a by-product.
- Light energy is converted into chemical energy by photosynthesis. **90%** of total photosynthesis is carried out by aquatic plants (**85% algae**) and 10% by land plants.
- First true and oxygenic photosynthesis started in cyanobacteria (BGA). Roots of tinospora and trapa are assimilatory or photosynthetic.
- **Absorption spectrum** of photosynthesis is blue and red light while **action spectrum** of photosynthesis is red & blue. While rate of photosynthesis is higher in red wavelength of light but highest in white light.
- **Blackman** discovered dark reaction while Calvin & Benson gave cyclic pathway for dark reaction. Dark reaction is also called as **Calvin cycle or C_3-cycle**.

- Q_{10} for dark reaction is between 2-3 while for light reaction is it is one.
- **Emerson & Arnold** worked on chlorella and gave the concept of two photosystem.
- When Emerson & Arnold gave only light greater than **680 nm** wavelength then quantum yield suddenly dropped down this is called as **red drop**.
- When Emerson & Arnold gave light shorter or greater than 680 nm wavelength then quantum yield increases, this is called as **Emerson effect or enhancement effect**.
- The number of O_2 molecule evolved by one quantum of light in photosynthesis is called as **Quantum yield**. The value of quantum yield is 1.25 or 12.5%.
- Quantasome are made up of 230-400 molecules of various pigments. Quantasome was discovered by **Park & Biggins**.
- Chlorophyll-a and carotene are **universal pigments** which are found in all O_2 liberating cells.
- 400-700 nanometre (nm) light used in photosynthesis is also known as **photosynthetic active radiation (PAR)**.

Pigments	**Formula**	**Found in**
Chlorophyll-a	$C_{55}H_{72}O_5N_4Mg$	All green plants
Chlorophyll-b	$C_{55}H_{70}O_6N_4Mg$	All green plants (except BGA,Red & brown algae)
Chlorophyll-c	$C_{35}H_{32}O_5N_4Mg$	Brown algae
Chlorophyll-d	$C_{55}H_{70}O_6N_4Mg$	Red algae
Carotene	$C_{40}H_{56}$	All green plants

Light reaction	
Photophosphorylation Cyclic ETS & Photophosphorylation	**Z-scheme or Non-cyclic ETS**
Only photosystem (PS)-I works	Both photosystem (PS)-I & PS-II works
Occurs in grana thylakoids & stroma thylakoids	Only grana thylakoids
No O_2 evolved	Oxygen (O_2) evolved
First electron acceptor is FRS. (Ferredoxin Reducing Substance)	First electron acceptor is Pheophytin and final electron acceptor is $NADP^+$

- Inhibitory effect of high conc. of O_2 on photosynthesis is called as **Warburg effect.**
- The inhibition of anaerobic respiration by O_2 concentration is called as **Pasteur effect.**
- 6 turns of Calvin cycles are required for the formation of one glucose.
- Comparison between C_3, C_4 and CAM pathway:

C_3-pathway	C_4-pathway	CAM-pathway
First stable compound is-3 carbon PGA	First stable compound is-4 carbon OAA	First stable compound is-4 carbon OAA
18 ATP & 12 $NADPH_2$ are used for 1 glucose form.	30 ATP & 12 $NADPH_2$ are used for 1 glucose form.	30 ATP & 12 $NADPH_2$ are used for 1 glucose form.
Kranz anatomy absent	Kranz anatomy present	Kranz anatomy absent
One type of carboxylase enzyme-Rubisco present	Two types of carboxylase enzyme- Rubisco & PEP case present	Two types of carboxylase enzyme- Rubisco & PEP case present
CO_2 acceptor is RUBP	Primary CO_2 acceptor are-RUBP & PEP case	Primary CO_2 acceptor are RUBP & PEP case
High CO_2 compensation point (40-100PPM)	Low CO_2 compensation point (8-10PPM)	High CO_2 compensation point (40-100PPM)
Transpiration ratio is 500-1000	Transpiration ratio is 200-300	Transpiration ratio is 50-100
Examples-mostly belongs to dicot family but wheat & rice are C_3-plants.	Examples-mostly monocot belongs to family-Gramineae & Cyperaceae. Such as-Sugarcane, Maize, Some dicot-Euphorbia, triplex, Amaranthus, Portulaca, Rosea, Tribulus & Chenopodium	Examples-Family Crassulaceae-kalanchoe, Bryophyllum, Sedum, Kleinia, opuntia, Crassula, Agave, Aloe, Pineapple & Welwistchia etc.
Discoverer- Calvin	Discoverer- Hatch & Slack	Discoverer- Oleary & Rouhani
Known as- Dark reaction, Blackman Reaction, Calvin cycle,C_3-Cycle, Biochemical phase, Carbon assimilation, Photosynthetic carbon reduction cycle (PCR-cycle),Reductive pentose phosphates pathway etc.	Known as- CO_2 conc. mechanism, Dicarboxylic acid cycle (DCA-cycle), C_4-cycle, Hatch & Slack pathway etc.	Known as- Dark CO_2 fixation,Crassulacean acid metabolism (CAM) and Dark acidification etc.

- Dimorphic chloroplasts are present in the leaves of C_4-plants. Chloroplast of bundle sheath (BS) cells or **Kranz cell** are larger & without grana while mesophyll chloroplasts are small & with grana. **Rubisco (Ribulose bisphosphate carboxylase oxygenase**) is present in BS cell while PEP in mesophyll cells. This type of anatomy is called **Kranz anatomy**.
- In C_4-plants, C_3-Cycle occurs in BS cells while C_4-Cycle occurs in mesophyll cells. First carboxylation in C_4-cycle occurs by **PEP (Phosphoenol Pyruvate) case** in mesophyll cytoplasm while final CO_2-fixation by C_3-cycle in bundle sheath cells.
- **Decker & Tio** discovered photorespiration and clarified that C_2-Cycle or glycolate pathway operates during only day time in C_3-plants & Rubisco act as oxygenase at higher concentration of O_2 & low CO_2 concentration in the C_3-green cells.
- Photorespiration occurs in chloroplast, mitochondria and peroxisomes. It is also called as three cell organelle reaction. It is found in **C_3-plants** & absent in C_4-plants.
- Factors affecting photosynthesis are light, temperature, CO_2, oxygen, water, chlorophyll, leaf age & orientation, inhibitors and minerals.
- Intensity of light at which rate of photosynthesis becomes equal to rate of respiration in plants is known as light compensation point. Net photosynthesis or **net primary productivity** at this point is **zero**.
- Optimum temperature for photosynthesis in plants is 20-35^{0}C.
- The amount of CO_2 in grams absorbed by 1 gram of chlorophyll in 1 hour is called as **photosynthetic number** or assimilatory number. This was given by **Willstatter & Stoll**.
- Plants which are adapted to grow in high intensity of light are called **heliophytes** while plants which are adapted to grow in shade are called **sciophytes.**
- **Law of minimum** was given by **Liebig**. According to it, when a process is governed by a number of separate factors, then the rate of process is controlled by that factors which is present in minimum amount.
- **Law of limiting factors** was given by **Blackman**. It is the modification of Law of minimum given by Liebig. According to it, "when a process is controlled to its rapidity by a number of factors, then the rate of process is limited by the pace of the slowest factors."

- Chlorobium, Chromatium, Rhodospirillum and Rhodopseudomonas are photosynthetic bacteria, having only pigment system, **PS-** . Only one ATP is produced in each turn of cyclic photophosphorylation & O_2 is not evolved.
- **Olson** in 1970 gave a non-cyclic scheme in bacterial photosynthesis.
- Pigment system of bacteria is denoted by **B-890 or 870**.
- The plants which are grown in N_2-deficient places (swampy place) but have insectivorous habits are called as insectivorous plants. Such as *Nepenthes* (pitcher plant), *Utricularia* (bladderwort), *Drosera* (sundew), *Pinguicula* (butterwort), *Dionaea* (Venus fly trap), *Aldrovanda* (water fly trap) and *Genlisa* (lobster pot trap).
- Non green plants which are depends on dead organic matter are called as saprophytes. Such as **Monotrapa** (Indian pipe) and **Neottia** (birds eye nest).
- Efficiency of photosynthesis is **30%** while efficiency of respiration is **42%**.
- **Respiration** is amphibolic & exergonic cellular process.
- When respiratory substrates are carbohydrates then respiration is known as **floating respiration**. If respiratory substrates are proteins then respiration is known as protoplasmic respiration.

Respiration	Oxidation (food)	Final products	ATP formed
Aerobic respiration	Complete	CO_2 & H_2O	38 or 36
Anaerobic respiration	Incomplete	C_2H_5OH & CO_2	2
Fermentation	Incomplete (Extracellular)	C_2H_5OH & CO_2	No ATP

- In aerobic respiration final electron acceptor is free oxygen. Anaerobic respiration was first reported by **Kostytchev**.
- **Glycolysis** occurs in cytosol or cytoplasm. It is common phase for aerobic & anaerobic respiration. Neither consumption of oxygen nor liberation of CO_2 take place.
- In glycolysis 1 glucose produces 2 mole of pyruvic acid (3C). 2NADPH$_2$ (6 ATP) & 2 A TP are produced in glycolysis, which are equal to 8 ATP.
- When the substrate releases energy for phosphorylation of ADP or formation of ATP without ETS then it is called as **substrate level phosphorylation**. Substrate level phosphorylation forms 4 ATP.

- Glycolysis is also known as Embden, Meyerhof and Parnas (EMP) pathway.
- Acetyl Co-A is a connecting link between glycolysis & Krebs-cycle. Acetyl Co-A is also common intermediate between fat & carbohydrate.
- Krebs-cycle was discovered by **H.A. Kreb**. It occurs in mitochondrial matrix. It is also called as **Tricarboxylic acid cycle (TCA) or Citric acid cycle**.
- All the enzyme of TCA cycle except **Succinic dehydrogenase** (inner mitochondrial membrane) are present in matrix.
- 3NADPH$_2$, 1FADH$_2$ & 1 GTP (ATP) are produced in each turn of TCA cycle. Which is equal to 12 ATP.
- Pentose phosphate pathway (PPP) is also called as Warburg-Dickens **pathway** or Hexose mono phosphate (HMP) Shunt, discovered by **Warburg & Dickens** and net gain is equals to 35 ATP.
- **Beta oxidation** of fatty acid takes place mainly in permitochondrial space but also in glyoxisome, peroxisome and cytosol.
- **Glyoxylate cycle** was discovered by Kornberg & Krebs during the germination of fatty seeds. Glyoxylate cycle occurs in glyoxisome, cytosol and mitochondria. This cycle converts fats into sugars so it is an example of **gluconeogenesis** in plants.
- **Cruick Shank &** Pasteur (1898) discovered the fermentation process. **Buchner** discovered the enzyme **zymase complex** which is responsible for alcoholic fermentation.
- The ratio of the volume of CO_2 released to the volume of O_2 taken in respiration is called as respiratory quotient (R.Q).

R.Q = volume of CO_2 liberated/ volume of O_2 consumed

Substrate	Value of R.Q
Carbohydrates	1
Fat or oil	0.7
Organic acid	More than 1
CAM Plants	0
Protein	0.9
Anaerobic respiration	∞ (infinite)

- Optimum temperature for respiration is **20-35°C**. The minimum amount of oxygen at which aerobic respiration take place & anaerobic respiration become extinct is called as **extinction point**.

- Oxygen concentration at which aerobic respiration & anaerobic respiration take place simultaneously is called as **transition point**.
- The rapid increase in rate of respiration during ripening of fruits, senescence of leaves and plant organs is called as **climacteric respiration**. It is due to production of ethylene.
- Paratonic movements in plants occur due to external stimulus. These movements are **directional**. Such as phototropism (stem), geotropism (root), chemotropism (pollen tube), thigmotropism (tendrils, Cuscuta) and hydrotropism (roots of seedlings).
- Nastic movements in plants occur due to external stimulus but diffused type. These movements are **non-directional**. Such as nyctinasty (flower, leaves & stomata daily movement), thigmonasty (tentacles of insectivorous plants), chemonasty (tentacles of insectivorous plants) and seismonasty (*Mimosa pudica*).

CHAPTER 10

Rangeland Management

- The total livestock population consisting of Cattle, Buffalo, Sheep, Goat, Pig, Horses & Ponies, Mules, Donkeys, Camels, Mithun and Yak in the country is 512.05 million in 2012. The total livestock population has decreased by about 3.33% over the previous census. It was **19th Livestock Census** of India.
- India supports about **15%** of the total livestock of the world. In India the area under permanent pasture is **12.47 million ha** which is about 4.1% of the total geographical area of the country.
- The grassland type and their associated environment are given in table:

Grassland type	Environment/region
Sehima/Dichanthium	Black soil
Dichanthium/Cenchrus	Sandy loams
Phragmites/Saccharum	Marshy area
Bothriochola	Paddy tracts & high rainfall belts
Cymbopogon	Low hills
Arundinella	High mountains
Deyeuxia/ Arundinella	Mixed temperate climate
Deschampsia/Deyeuxia	Temperate Alpine climate

- **Dabadghao and Shankar Narayan** (1973) have recognised five types of grass covers in India. These are given in the table:

Grass cover type	Found in	Covered area	Soil type
Sehima/Dichanthium	Tropical area	1.74 million Sq. km	Red,Black,Yellow & laterite soil
Dichanthium-Cenchrus-Lasiurus	Sub-tropical & semi-arid region	4 lakh Sq. km	Alluvial & Grey brown soil

Phragmites-Saccharum-Imperata	Gangetic plan & Brahmaputra valley	2.8 million Sq. km	Alluvial soil
Themeda- Arundinella	Hilly region	2.3 lakh Sq. km	-
Temperate-Alpine	High hilly region (Northern)	-	-

- CAZRI Range condition Classification system is given below:

Range condition Class	Air dry forage yield	Carrying capacity(ACU/1000)
Excellent	$1^{1/2}$ tons per ha	25-30
Good	1 ton per ha	20
Fair	¾ ton per ha	17
Poor	½ ton per ha	13
Very poor	200 kg per ha	-

- **Carrying capacity** refers to "the number of individual animals that can survive the greatest period of stress each year on a land area". Carrying capacity is the number of grazing animals a piece of land can support for a long term without causing harm to rangeland resources (vegetation, soils, and water).
- **Stocking rate** refers to the number of animal grazing a unit of area (usually 1 ha) at a particular time.
- **Livestock unit** refers to the feed requirement used as the basis of comparison for different classes and species of stock.
- 1 livestock unit requires approximately **520 kg** of good quality pasture dry matter per year.

Cattle	Equivalent cow unit
Cow or Bullocks	1
Buffalo	2
Camel	8
Goat	½
Sheep	½
Other animal	1

- 1 animal unit (au) is equal to **40.0 kg** steer. Breeding cow and bull have **2.0 au**.

- The **grazing capacity per ha** on animal basis can be calculated as:

$$\text{Grazing capacity (GC)} = \frac{\text{Total annual forage production.....}}{\text{Forage requirement per animal} \times 365 \times a}$$

Where '*a*' is the area of grazing land.

- 100 cows and 20 buffalos with a planned grazing schedule from May 1 – July 1 and 4 camel grazed from July 1 – November 1. Calculate total cow unit (per month) of all grazing animal.

100 cows = 100 × 1 cow unit (for 2 month [100×2] = **200**)

20 buffalos = 20× 2 cow unit (for 2 month [40×2] = **80**)

4 camel = 4 × 8 cow unit (for 4 month [32×4] = **128**)

Total cow unit = 172 cow unit = **408 per month.**

- An area of **1,000 ha** of forest with grazing capacity of **0.8 ha** per cow unit is to be managed under rotational grazing. Calculate how many cow units can graze.

Total grazing area = 1,000

Grazing capacity = 0.8 ha/cow unit

$$\text{Then total cow unit} = \frac{\text{Total grazing area}}{\text{Grazing capacity}} = \frac{1{,}000}{0.8} = 1250 \text{ cow units.}$$

- Minimum seed rate for some species are given below:

Species	Seed rate (kg) per ha
Lasiurus sindicus	4.5
Cenchrus ciliaris	4.5
Panicum antidotale	5.6
Dichanthium annulatum	4.5
Pennisetum pedicellatum	8.9
Sehima nervosum	7.8

- The object of **scientific management** of grassland is to maintain the grassland in the highest state of production.
- When the number of animals that are allowed to graze per unit area of rangeland is fixed accordance with rangeland carrying capacity and grazing is not allowed during plants are passing through the critical stage of growth it is called **controlled grazing**.

- When the animals are kept permanently on a given area of a rangeland and allowed to move freely over this area, due to this the crop composition is changed. This is called **continuous controlled grazing**. This is the most common practice of grazing in India.
- When the grassland is divided into three **compartments** and the grazing being allowed in two compartments while third one is given rest it is called **deferred grazing**. This system has been found more suitable for perennial grasses.
- When the grassland is divided into two or more **blocks** and the grazing is allowed in one of the blocks while other block gets rest and next year cattle are moved to other block to give rest to the previous block is called **rotational grazing**.
- **Deferred rotational grazing** is a combination of deferred and rotational grazing. In this system grazing is allowed by dividing the grazing area into two block A and B. The grazing is allowed in block A in the first half of the season and the block B is given rest for seed maturity. The block B is opened for grazing during the last half of the grazing season. After a lapse of two year the original sequence is repeated. This system has increased the grazing capacity upto 39%.
- The conservation of green forage into dry form without affecting the quality of the original material is called **hay**. The moisture content in hay is upto 15%.
- The product obtained by packing fresh fodder in a suitable container and allowing it to ferment under anaerobic condition without undergoing much loss of nutrients is called **silage**.
- The process of preserving green fodder in a suitable container and allowing it to ferment under anaerobic condition is called **ensilage**. Silage is prepared by this method after 2-3 months.
- Some important grasses with their common names:

Common name	Botanical name	Average forage yield
Buffalo grass or Paragrass	*Brachiaria mutica*	2000-2500 Qt./ha
Anjan grass	*Cenchrus ciliaris*	300-350 Qt./ha
Bermuda grass or Doob grass	*Cynodon dactylon*	300-350 Qt./ha
Sain grass	*Sehima nervosum*	250-300 Qt./ha
Dinanath grass	*Pennisetum pedicellatum*	750-1000 Qt./ha
Jarga grass	*Dichanthium annulatum*	200-250 Qt./ha

Barware grass	*Panicum antidotale*	200-300 Qt./ha
Guinea grass	*Panicum maximum*	750-1600 Qt./ha
Sudan grass	*Sorghum sudanense*	260-300 Qt./ha
Napier grass, elephant grass	*Pennisetum purpureum*	40 ton/ha
Tiger Grass	*Thysanolaena maxima*	4 ton/ha

- Duration of important census in India is given below:

Name	Duration (In years)
Livestock population census	5
Population census	10
Wildlife population census	5
Forest survey report	2
Working plan	10

CHAPTER 11

Forest Protection

- **Forest protection** is a branch of forestry which deals with the activities directed towards the prevention and control of damage to forest by man, animals, fungi, insects, injurious plants and adverse climatic factors.
- According to **Swaminathan**, about 6000 million tonnes of top soil in India are washed down or blown away every year. Due to this there is about 6 million acre of cultivable (top 18 cm soil) land loss every year.
- According to **Wiackowski**, the leaf surface is 10 to 20 times greater than the earth's surface occupied by plants.
- Dense plantation of **Pitheolobium dulce** is very beneficial in reducing burning coal pollution.
- World average growing stock per hectare is **100 m^3** and m.a.i. is **2.0 m^3 per ha**. While, India's average growing stock per hectare is **32 m^3** and m.a.i. is **0.5 m^3 per ha**.
- Forest protection measures can be broadly classified into two categories viz. first is **preventive measures** and second is **remedial measures**.
- Preventive measures are those measures which prevent occurrence of damage to forest from damage causal agency.
- The remedial measures are those measure operations which are carried out to minimize the damage after damage has occurred.
- **Deforestation** refers to illicit felling of trees from a piece of land without the intension of reforestation.
- **Dr. Voelcker** submitted his report on "**Improvement of Indian Agriculture**" in 1893.
- **Encroachment** may be defined as an act of seizing prossession of some forest land by the people.

- A place where the timber or other forest produce required by villagers for their bonafide domestic and agricultural use are obtained is **called nistar (local depots)**.
- Forest fire are caused by **natural causes (5%)**, man's carelessness **(95%)** and man's deliberate & intentional actions.
- About 95% of the fire in India are caused by man. While fire due to natural causes is 5%.
- Table showing classification of forest fire:

A. On the basis of causative factors
1. Natural fires 2. Accidental fires 3. Deliberate or intentional fires
B. On the basis of the place of their action
1. Creeping fire 2. Ground fire 3. Surface fire 4. Crown fire

- **Creeping fire** is defined as a forest fire spreading slowly over the ground with low flame in the absence of strong wind. This fire occurs usually in summer nights and burn dry leaves cover.
- **Ground fire** is defined as a forest fire that burns ground cover (herbaceous plants and shrubs) only.
- **Surface fire** is defined as a forest fire which burns not only the ground cover but also undergrowth. This fire most commonly occurs in plains.
- **Crown fire** is defined as a forest fire which spreads through the crowns of trees and consumes all or part of the upper branches and foliage. This fire most commonly occurs in coniferous forests.
- Broad-leaved trees are less affected by fire as compare to conifers. Chirpine tree is fire resistance due to its thick corky bark.
- Weather, inflammable materials and topography factors make up the fire environment.
- High temperature and hot strong wind together prepare an environment in which the fire risk is maximum. But even in a favourable environment, fire occurs only when there is presence of inflammable materials on the forest ground.

- The unusable residue after logging and conversion operations is called as **slash**.
- Forest fire spreads fast uphill and slowly downhill side. On a slope, it spreads fast towards the ridge and slowly towards the valley. And on wind it assumes a circular shape.
- Speed of forest fire varies from 0.3 meter per minute to **107 meter per minute**. While fire speed under forest canopy does not exceed 6 meter per minute.
- Table showing season of forest fire:

Region	Fire season start from
Plains (Northern & Central India)	March to June
Sub mountain chirpine forest & deodar forests	May to June
Southern India	Mid-January to April ending

- Burning of leaves litter or other undergrowth before completely drying to prevent later fire damage is called **early burning or controlled burning**.
- **Slash disposal** is defined as the treatment or handling of slash for reducing hazards from fire, fungi or insects and for providing the seeds with access to soil.
- *Strobilanthus* act as a natural fire break in evergreen forests due to its juicy stem.
- **Firelines** is defined as a cleared permanent fire break intended to prevent fires from crossing from one area to another.
- Firelines are of two kinds viz. internal firelines and external firelines. Internal firelines (width 6m to 30 m) are set inside the forest to prevent the fire.
- In India, fire is usually extinguished by the following methods viz., by water, by air, by beating and by counterfiring.
- **A fire trace** is defined as a cleared lines used as a base from which to counterfire. Map prepared by using 1:50,000 R.F scale is called fire map.
- *Xyleutes ceramicus* (Beehole) is more abundant in teak plantation than in its natural forest.

- Table showing defoliators of some important tree species:

Tree species	Common name	Defoliator name
Tectona grandis	Teak	*Hyblaea puera*
Cedrus deodara	Deodar	*Ectropis deodarae*
Dalbergia sissoo	Shisham	*Plecoptera reflexa*
Tree species	**Common name**	**Shoot Borer name**
Toona ciliata	Toon	*Hypsipyla robusta*
Swietenia mahagoni	Mahogany	*Hypsipyla robusta*
Chukrasia tabularis	Chukrasia	*Hypsipyla robusta*
Bombax ceiba	Semal	*Tonica niviferana*

- *Ips longifolia* is a bark borer of *Pinus longifolia* (chir). *Hoplocerambyx spinicornis* is the **sal heartwood borer**.
- *Hapalia machaeralis* is teak **skeletonizer**.
- Cockchafers (root feeders), cutworms (cut seedlings) and crickets (seedling feeders) are major pest of forest nursery.
- *Lantana* and *Eupartium* are the examples of obnoxious weeds.
- Forest trees diseases can be classified into two classes viz, root diseases and heart rot disease. These are given in table:

Tree	Root rot	Heart rot	Wilt
Dalbergia sissoo	*Ganoderma lucidum*	–	*Fusarium solani*
Accacia catechu	*Ganoderma lucidum*	*Fomes badius*	–
Shorea robusta	*Polyporus shoreae*	*Hymenochaete rubiginosa*	–
Tectona grandis	*Polyporus zonalis & Peniophora rhizomorpha-sulphurea*	–	*Pseudomonas solanecearum*
Eucalyptus hybrid	*Ganoderma lucidum*	–	–
Casuarina equisetifolia	–	–	*Trichosporium vesiculosum*
Abies pindrew	*Armillarea mellea*	*Fomes* spp.	–
Picea smithiana	*Armillarea mellea*	*Fomes* spp.	–
Cedrus deodara	*Fomes annosus*		–

- The causal organism of Spike disease of sandal is **mycoplasma**. These are known as MLOo.
- **Corticium salmonicolour** cause pink disease in eucalyptus tree.
- **Cronartium himalaynese** commonly known as chirpine Swertia felt rust cause mortality in young **Pinus longifolia** (chir). While **Poria monticola** causes cuboidal decay in it.
- **Fomes pini** commonly known as **Trametes pini** causes decay in heartwood of blue pine.
- **Fusarium** develops red colour of wood in spruce and fir instead of white colour of wood.

CHAPTER 12

Silviculture

- The word **forest** is derived from Latin word "*foris*" meaning "outside the village boundary" or away from the inhabited land.
- **Silviculture** refers to certain aspect of theory and practice of raising forest crops. It is defined as the art and science of cultivating forest crops.
- **Silvics** has been defined as, the study of life history and general characteristics of forest tree crops with particular reference to environmental factors as the basis of the practice of silviculture.
- **Forest** is an area set aside for the production of timber and other forest produce or maintained under woody vegetation for certain indirect benefits which it provides.
- Classification of forest is shown in table:

Age	Regeneration	Composition	Ownership	Management	Growing stock
Even-aged	High forest	Pure forest	Public forest- 1. Reserved 2. Protected 3. Village	Protection forest	Normal forest
Uneven-aged	Coppice forest	Mixed forest	Private forest	Production forest Social forest	Abnormal forest

- **High forest** is regenerated from seed and coppice forest is regenerated from the vegetative parts of trees either naturally or artificially.
- **Pure forests** are composed almost entirely of one species (not less than 50%) while mixed forests are composed of two or more than two species.
- A **reserved forest** is an area with complete protection, constituted according to chapter II of Indian forest Act-1927.

- A **protected forest** is an area with limited degree of protection, constituted according to chapter IV of Indian forest Act-1927.
- A **village forest** is a stable forest assigned to a village community under the provisions of chapter III of Indian forest Act-1927.
- **Forestry** is defined as the theory and practice of all that constitutes the creation, conservation and scientific management of forests and the utilization of their resources.
- The foundation of scientific forestry was laid during **1864** with creation of forest department by Government of India and appointment of the first Inspector General of Forest, **Dr. Dietrich Brandis**.
- The Central Board of Forestry (CBF) was constituted in **1948** and **Van Mahotsav** (festival of tree planting) was started in **1950**.
- Out of total geographical area of **328.8 million hectare** in India, approximately **76 million hectare** land is classified as forests.
- About **95.8 per cent** of the forests of the country are owned by the Government and most of the forests in India are of tropical climate type.
- **Site** is defined as a complex of physical and biological factors of an area that determines what vegetation it may carry.
- **Site quality** is defined as the relative productive capacity of a site for particular species.
- The portion of the solar energy to which human eye is sensitive is known as **light**.
- The sky appears blue because of the **scattered** light which is blue in colour.
- The ability of a surface to reflect the solar radiation is called **albedo**.
- The species which require abundant light for their optimum growth and development are called **light demander**. A species which require shade atleast in early stages for their optimum growth and development are called **shade demander**.
- A species which are capable of persisting and developing under shade is called **shade bearer**.

Light demander	Shade bearer	Shade demander
Dipterocarpus, Calophylum, Shorea, Bombax, Adina, Pinus, Populus, Teak, Sissoo etc.	*Artocarpus, Pterocarpus, Gmelina, Quercus, Cedrus, Picea, Cupressus, Toona, Dalbergia latifolia, Pterocarpus*	*Mesua, Xylia, Schleichera, Syzygium, Taxus , Abies, Mallotus, Careya, Taxus*

- Chilling of air below the its freezing point is called **frost.**

Frost name	Occurrence	Produced by	Effects on
Radiation frost/ Ground frost	Clear sky nights in winters	Loss of heat by radiation	Seedlings killing in nursery
Pool frost/con-vection frost	Dehra Dun valley. It is common in the valley.	Accumulation of heavy cold air into natural depression	Large poles and trees. More inju-rious
Advective frost	During winters	Cold air brought from outside	Vegetation

- **Temperature** is perhaps one of the most important single factor which determins the distribution of vegetation.
- **Rainfall intensity** is the rate at which rain falls. A day in which 2.5 mm or more rain falls is called a rainy day.

Drought hardy species	Moderately hardy species	Drought sensitive species
Acacia nilotica, A. senegal Boswellia serrata, Hardwickia binata, Pongamia, Ougenia, Albizia, eucalyptus, Diospyrus, Azadirachta, Tecoma undulata	*Acacia catechu, Adina cordifolia, Albizia procera, Dalbergia sissoo, Emblica officinalis, Mitragyna parviflora, Dendrocalamus strictus,*	*Madhuca indica, Mangifera indica, Pterocarpus marsupium, Shorea robusta, Tectona grandis, Terminalia species, Toona ciliata.*

- *Potential productivity* means a productivity which is really obtainable of at a given condition.
- CVP (climate, vegetation and productivity) index was given by **Paterson** in 1956.
- Altitude refers to the height of a place from mean sea level. **Slope or gradient** is defined as the angle formed by the surface of the land with the horizontal plane.
- **Aspect** is defined as the direction towards which a slope faces or it is the direction of slope of a land.
- **Epiphytes** are plants which grow on the trunk and branches of other plants for the purpose of support. Important epiphytes are banyan, pipal and orchids.
- Most diseases in plants are caused by Thallophytes (fungi, bacteria and lichens).

Disease name	Causal organism
Damping-off	*Rhizoctonia, Pythium and Fusarium.*
Heart-rot	*Fomes fungi*
Wilt disease	*Fusarium fungi*
Root disease	*Ganoderma* spp.
Pink disease (Eucalyptus)	*Corticium salmonicolor*
Spike disease (Sandal tree)	Mycoplasma (PMLOs)

- Development of soil also depends upon the climate to a great extent in following table:

Climate (Area)	Soil formation
Hot & dry area	Desert soil with high per cent of sand & salts
Hot-humid & hot wet	Grey-brown, red and lateritic soil
Colder regions	Podsol soil

- Table showing forms and shape of crown

Crown shapes	Examples
Conical	Pine & Deodar
Cylindrical	Fir, Spruce and Eucalyptus
Spherical	Mango, Neem, Imli, Mahua
Broad & flat topped	*Acacia planifrons*
Broom	Babul
Frondose	*Prosopis juliflora*

- Tree with single apical meristem are termed as **monoaxial trees**.
- **Forking** is common in broad leaved species and it is caused due to injuries in the growing apical meristem. While buttressing is common in wet and moist tropical forests.
- When a stem shows irregular involution and swellings it is called **fluting**.
- The symbiotic association between certain fungi and roots of higher plants which enhances the uptake of water and nutrients is called **mychorrhiza**.
- The branch of botany which deals with the study of internal structures and organization of plants or plant organs is called **Plant Anatomy**.

- From germination to the stage where the young trees has a few leaves called **recruit**.
- From recruit stage upto a height of 1 m of the young plant it is called **seedling**.
- From the time the young tree reaches 1 m height till the lower branches begin to fall and having dead bark on the stem is called **sapling**.
- From the fall of the lower branches to the time when the rate of increase in height begins to fall and crown expansion becomes marked it is called **pole**.
- **Phenology** refers to the seasonal changes in the development of foliage, flowering and fruiting. The peak growth period may depend upon species and locality but it is usually during June to September.
- **Annual growth rings** are generally used for finding out the age of the tree.

Distinct growth ring species	**Not distinct growth ring species**
Pinus spp., *Cedrus, Abies, Acacia catechu, Anthocephalus, Gmelina, Betula, Artocarpus, Toona, Lagerstroemia, Michelia and Tectona* etc.	*Adina, Albizia, Dalbergia, Bombax, Sterculia, Schima wallichii and Trewia nudifera* etc.
Ring porous wood species	**Semi-ring porous wood species**
Tectona, Morus, Melia and Dalbergia.	*Juglans, Toona and Pterocarpus.*

- The operations which are carried out in the forest crops from seedlings to mature stages to provide a healthy environment for their growth and development is called **tending operations**.
- Tending operations includes weeding, cleaning, thinning, improvement fellings, runing and **climber cutting**.
- **Weeding** is defined as removal of all unwanted plants, particularly in seedling stage at nursery or in forest crops to reduce the competition for moisture, nutrients and to provide sufficient growing space for the desired species.
- **Lantana camera** is a notorious weed controlled by *Orthezia insignis* (lantana bug).
- **Cleaning** is carried out in sampling stage.
- **Thinning** may be defined as a felling made in immature stand for the purpose of improving the growth and form of the trees that remains without permanently breaking the canopy.

- Thinning may be an essential tool for shortening the rotation of a forest crop.
- Kinds of thinnings used in regular crops are mechanical thinning, ordinary thinning, crown thinning, free thinning, Craib advance thinning and numerical thinning.
- Thumb rule for thinning is $d = aD + b$. where, a & b are constant, d = spacing (ft), D = average diameter in inches.
- Some commonly used formulae in thinning are given below:

Formula	Used for	Given by
$d = D$	Deodar	Glover
$d = 1.5\ D$	Irregular sal crops	Laurie
$d = 2\ D$	Sissoo	Sagreiya
$d = 3\ D$	Irregular teak crops	Howard

- **Ordinary thinning** is also called German thinning or low thinning or thinning from the below. In ordinary thinning trees are removed from lower classes and it is most commonly used practice in forestry. Most suited for light demander species.
- Ordinary thinning is classified into five grades as given in the table:

Light thinning (A)	Moderate thinning (B)	Heavy thinning (C)	Very Heavy thinning (D)	Extremely Heavy thinning (E)
Removal of dead, dying, diseased and suppressed trees of classes-III, IV and V.	Removal of dead, dying, diseased and suppressed and an occasional very defective dominant (c) trees of classes-I (d), II (b), III, IV and V.	All classes of trees of B grades and removal of remaining dominated and defective codominants without making lasting gaps in canopy. Classes-I (b), I (c), I (d) II, III, IV and V.	All classes of trees of C grades and some goods dominants of some I (a) without making lasting gaps in canopy. Classes-I (b), I (c), I (d) II, III, IV and V.	All classes of trees of D grades and more dominants of (a) without making permanent gaps in canopy. Classes-I (a), I (b), I (c), I (d) II, III, IV and V.

- Table showing coppicing capacity of forest species:

Strong coppicers	Fair coppicers	Poor coppicers	Non- coppicers
Acacia catechu, Albizia spp., *Anogeissus latifolia, Butea monosperma, Dalbergia sissoo, Diospyrus melanoxylon, Ougenia oojensis, Shorea robusta, Tectona grandis, Eucalyptus* spp., *Salix* spp.	*Pterocarpus marsupium, Terminalia* spp., *Quercus leucotrichophora, Hardwickia binata, Manilkara hexandra*	*Acacia nilotica, adina cordifolia, Madhuca indica, Bombax ceiba*	*Pinus* spp., *Cedrus deodara, Picea smithiana, Abies pindrew*

- **Forest Nursery** is an area where plants are raised for eventual planting out has ordinarily both seedlings and transplants.
- **Seedling Nursery** is a nursery which has only seedling beds for raising seedlings.
- **Transplant Nursery** is a nursery which has only transplant beds for transplanting seedlings.
- For coniferous species acidic soils having **pH 5 – 6.5** is reported to be good for their growth.
- The size of nursery varies from 0.5 to 2.5 per cent of total plantation area.
- Distance between bed to bed in a forest nursery is **45 cm.**
- Normal bed size is **10 m × 1 m** while standard bed size is **12.5 m × 1.2 m** in forest nursery.
- **Area of nursery** $(A) = \dfrac{18 \times 1.2 \times X \times Y}{z}$

  where X = plantation area (ha). Y = required plant/ha. z = number of plant in a bed.
- As a rule rectangular shaped beds are preferred than other shapes.

Name	Height/depth	Climatic conditions	Species
Raised beds	10 – 15 cm above ground level	Heavy rainfall area	Teak
Sunken beds	10 – 30 cm deep	Dry area	–
Level beds	Surface level	Normal rainfall area	–

- The soil in beds should be dug out to a depth of **40 to 50 cm**.
- Ratio of Sand: Soil: FYM should be **1 : 3 : 1** for soil filling polybags in nursery.

Polythene bag size (in cm)	Uses for growing or germination
10 × 20	Seed germination
15 × 20	month seedling
20 × 30	8– 12 month plant growing in nursery

Sowing method	Seed size	Suitable for	Seed required
Broadcasting	Large	Small area	3 – 4 times more
Dibbling sowing	Large sized	Small area	Equal amount
Drill sowing	Small and large seeds	Large area	2 times more

- Depth of seed sowing according to seed diameter:

Soil Depth	Seed size
Diameter of seed	Medium or bigger
2.5 to 5mm	Minute

- **Quantity of seeds** (W) = $\frac{A \times D}{P \times N} \times 100$

  where, W = required seed wt. (gm.)

  A = seed bed area (m^2), N = number of seed/gm.

  P = plant per cent of seeds, D = number of plants required foe sq. meter.

- Best time of seed sowing in a nursery is **March to June**.
- Stump planting is common practice in **Teak** and **Sissoo**.
- 1 year old seedling of teak and sissoo having tap root length not less than 30 cm and attaining thumb thickness (1.5-2.0 cm) is suitable for stump planting.
- **Transplanting** is moving of plant from one nursery bed to another for better root and shoot growth. This process is also called as **pricking out or lining out.**
- The plants which have been pricked out in the nursery are called **transplants**.
- **IBA** is a root promoting hormone in cuttings.
- A month which has less than **50 mm** of rainfall is usually referred to as dry month.
- **Provenance** refers to area where the mother trees of the seeds are growing.

- **Seed viability** in normal storage conditions:

Viability Period	Species
Two weeks	*Dipterocarpus* spp., *Shorea, Hopea, Michelia, Azadirachta etc.*
Three months	*Chloroxylon* spp., *Mangifera, Podocarpus* spp. *etc.*
One year	*Artocarpus, Toona, Mesua, Tectona, Terminalia* spp. *etc.*
Two or more than two years	*Accacia* spp., *Albizia* spp., *Cassia* spp., *Prosopis* spp., *Ochroma* spp., *Leucaena leucocephala and Caesalpinia* spp. *etc.*

- **Seeds per kilogram** is determined by counting five replicates of 100 seeds. While viability test required four replicates of 100 seeds.
- **Germinative capacity** means the total number of seeds that germinate in the test plus the number of sound seeds remaining ungerminated at the end of the test.
- **Germinative energy** is the percentage of seed in a sample that have germinated in a test upto the time when the rate of germination reaches its peak.
- **Tetrazolium test** is performed to test seed viability. This is based on reduction process which takes place in living cells. The indicator used is a colourless aqueous solution (1%) of **2, 3, 5- triphenyl-tetrazolium chloride/bromide**. By hydrogenation, this indicator is transformed into a red, stable and non-diffusible triphenyl-formazan in living cells.
- Table showing types of pits for sowing plants:

Pits type	Soil type	Climatic conditions
Ordinary pits	Clayey & clayey loam	All rainfall classes
Saucer shaped pits	Sandy	Medium rainfall
Ring pits	Very sandy	Less rainfall

- Marking of planting spots or soil-working spots with the help of sticks is called **staking**.
- **Seed rate** is the requirement of seeds for sowing a hectare of area.
- 50-100 cm height of planting stock is considered good for most of the species in India. While **1 : 2 ratio** of root: shoot is consider good for several species.

- The distance between plant to plant in a plantation or in a crop is called **spacing**.
- Table showing methods of calculating **number of plants** (N):

Planting pattern	Formula	Remarks
Line planting	$N = \dfrac{100\times100}{X\times Y}$	Where, X = distance of plants in lines Y = distance between the lines
Square planting	$N = \dfrac{100\times100}{(X)^2}$	X = Square of planting distance.
Triangular planting	$N = \dfrac{100\times100\times1.155}{(X)^2}$	X = Square of planting distance.
Quincunx planting	$N = \dfrac{2\times100\times100}{(X)^2}$	X = Square of the side of planting square.

- **Singling** is an operation under which forked or multiple stems are reduced to a single stem to improve the form of the planted tree.
- **Re-spacing** is an operation in which competing plants of the same or similar species are removed to provide proper growing space.
- Table showing standard parameters of some fencing:

Fencing name	Specifications
Trench fencing	Depth = 1.25 meter, top width = 1.25 m wide, bottom width = 0.75 m
Stone wall fencing	1.25 m in height
Wire fencing: (total ht. = 1.25 m) With four strands	1st strand height from ground = 15 cm 2nd strand height from ground = 45 cm 3rd strand height from ground = 60 cm 4th strand height from ground = 75 cm

- An **indigenous species** is one of that grows naturally in the country or in a region.
- An **exotic species** is one which is grown outside the limits of its natural range.

- List of successful exotics in India:

Species	Place of introduction	Native/origin country
Acacia auriculiformis	FRI, West Bengal, Bihar	Queensland, Australia
Acacia tortilis	Jodhpur (Rajasthan)	Israel
Casuarina equisetifolia	Tamil Nadu	Indonesia
Cinnamomum camphora	West Bengal	China
Cryptomeria japonica	Dehra Dun	Japan
Dendrocalamus giganteus	Rajasthan	Burma
Eucalyptus alba	Dehra Dun	Indonesia
Eucalyptus globulus	Nilgiris	Australia
Fraxinus excelsa	Kulu (H.P)	Europe
Hevea brasiliensis	Kerala	Brazil
Grevillea pteridifolia	Amarkantak (M.P)	Australia
Leucaena leucocephala	Dehra Dun, Gujarat	Hawai, Peru
Pinus nigra	West Bengal, H.P	Europe
Populus deltoides	Haryana, Punjab	British Forestry Commission
Prosopis juliflora	Rajasthan	Mexico
Prunus armeniaca	Kashmir, H.P	China
Pseudotsuga taxifolia	H.P	North America
Salix alba	Kashmir, H.P	China
Swietenia macrophylla	West Bengal, Karnataka	Central America
Swietenia mahogoani	Maharashtra	Cuba, Jamaica

- **Champion and Seth** (1968) classified Indian forests into 5 major groups and 16 type groups:

Five Major groups	Sixteen Type groups
I. Tropical forest mean annual temp (mat) : > 24° C	1. Wet evergreen forests 2. Semi- evergreen forests 3. Moist deciduous forests 4. Littoral and swampy forests 5. Dry deciduous forests 6. Thorn forests 7. Dry evergreen forests

II. Montane sub-tropical forests mat : 17° - 24° C	8. Subtropical broad leaved forests 9. Subtropical pine forests 10. Subtropical dry evergreen forests
III. Montane temperate forests mat : 7° - 17° C	11. Montane wet temperate forests 12. Himalayan moist forests 13. Himalayan dry forests
IV. Sub-alpine forests (mat- < 7°)	14. Sub-alpine forests
V. Alpine scrub	15. Moist alpine
	16. Dry alpine

Forest type	Mean Annual rainfall (mm)
Wet forest	Mean annual rainfall – > 2500 mm
Moist and semi moist forests	Mean annual rainfall – 1250 - 2500 mm
Dry forest	Mean annual rainfall – 750 - 1250 mm
Desert forests	Mean annual rainfall – < 750 mm

- Table showing flowering time, seed collection and master seed year of important trees:

Species name	Time of flowering	Seed collection time	Master seed year
Abies pindrew	-	October-Nov	10
Acacia catechu	June-Sept.	Jan-Feb	1-2
Adina cordifolia	June-Aug.	April-June	–
Ailanthus excelsa	Feb-Mar.	April-May	–
Albizia lebbeck	Sept-Nov.	Jan-Feb.	–
Albizia procera	Sept-Nov.	Feb-Mar.	–
Anogeissus latifolia	June-Sept.	Dec-Mar.	–
Azadirachta indica	Mar.-May	June-July	–
Bombax ceiba	Jan-Mar.	April-May	–
Boswellia serrata	Feb-April	May-June	–
Casuarina equisetifolia	May	June-Dec.	–
Cedrus deodara	–	Oct.-Nov	4-5
Cupressus torulosa	–	Oct.-Nov	7-8
Dalbergia sissoo	April-May	Dec-Feb	1-2
Dalbergia latifolia	Aug.-Sept.	Jan-Feb	–

Diospyros melanoxylon	Feb-April	May-June	Alternate year
Dendrocalamus strictus	Mar.-April	May-June	–
Eucalyptus hybrid	June-July	Dec-Jan.	Every year
Gmelina arborea	Nov-Dec	May-June	–
Juglans regia	April-May	Sept-November	–
Morus alba	Oct-Feb	April-May	–
Picea smithiana	–	Oct.-Nov	5-6
Pinus roxburghii	–	Mar.-April	4-5
Pinus wallichiana	–	Sept-Oct.	2-3
Populus ciliata	–	-	
Prosopis juliflora	Feb-March	May-June	Every year
Pterocarpus marsupium	Sept-Oct.	Dec-April	–
Quercus species	April-June	Aug-Sept	–
Santalum album	Sept-Dec	Mar.-April	–
Shorea robusta	March-April	June-July	3-5
Tectona grandis	July-August	Nov-Jan.	–
Toona ciliata	Feb-April	May-June	Alternate year

- Table showing silvicultural characteristics of important Indian trees (Y = Yes):

Species name	**Drought hardy**	**Moderately Drought hardy**	**Drought sensitive**	**Frost hardy**	**Frost tender**
Abies pindrew	-	-	Y	-	Y
Acacia catechu	-	Y	-	Y	-
Acacia arabica	Y	-	-	-	Y
Adina cordifolia	-	Y	-	-	Y
Ailanthus excelsa	Y	-	-	-	Y
Albizia lebbeck	-	-	Y	-	Y
Albizia procera	-	Y	-	-	Y
Anogeissus latifolia	-	-	Y	Y	-
Anogeissus pendula	-	Y	-	Y	-
Azadirachta indica	Y	-	-	-	Y
Bombax ceiba	Y	-	-	-	Y

Boswellia serrata	Y	-	-	-	Y
Casuarina equisetifolia	-	-	Y	-	Y
Cedrus deodara	-	-	Y	Y	-
Cupressus torulosa	-	-	Y	Y	-
Dalbergia sissoo	-	Y	-	Y	-
Dalbergia latifolia	Y	-	-	Y	-
Diospyros melanoxylon	Y	-	-	Y	-
Dendrocalamus strictus	-	-	Y	Y	-
Eucalyptus hybrid	-	-	Y	-	Y
Hardwickia binata	Y	-	-	Y	-
Gmelina arborea	-	Y	-	Y	-
Juglans regia	-	-	Y	-	Y
Madhuca indica	-	-	Y	Y	-
Mallotus philippinensis	Y	-	-	Y	-
Morus alba	-	-	Y	-	Y
Mitragyna parviflora	-	Y	-	-	Y
Picea smithiana	-	-	Y	-	Y
Pinus roxburghii	-	-	Y	Y	-
Pinus wallichiana	-	-	Y	Y	-
Populus ciliata	-	-	Y	-	Y
Prosopis juliflora	Y	-	-	-	Y
Pterocarpus marsupium	-	-	Y	-	Y
Quercus species	-	-	Y	Y	-
Schleichera oleosa	Y	-	-	Y	-
Shorea robusta	-	-	Y	Y	-
Syzygium cumini	Y	-	-	-	Y
Tectona grandis	-	-	Y	-	Y
Terminalia spp.	-	-	Y	-	Y
Toona ciliata	-	-	Y	Y	-
Ulmus wallichiana	-	-	Y	Y	-

- Table showing silvicultural characteristics of important Indian trees (Y=Yes):

Species name	Light demander	Shade demander	Shade bearer
Abies pindrew	-	Y	-
Acacia catechu	Y	-	-
Acrocarpus fraxinifolius	Y	-	-
Adina cordifolia	Y	-	-
Ailanthus excelsa	Y	-	-
Albizia lebbeck	-	-	Y
Albizia procera	-	-	Y
Anogeissus latifolia	Y	-	-
Azadirachta indica	Y	-	-
Bombax ceiba	Y	-	-
Boswellia serrata	Y	-	-
Casuarina equisetifolia	Y	-	-
Cedrus deodara	-	-	Y
Cupressus torulosa	-	-	Y
Dalbergia sissoo	Y	-	-
Dalbergia latifolia	-	-	Y
Diospyrus melanoxylon	Y	-	-
Dipterocarpus spp.	Y	-	-
Dendrocalamus strictus	-	-	Y
Eucalyptus hybrid	Y	-	-
Gmelina arborea	Y	-	-
Juglans regia	-	-	Y
Mallotus philippinensis	-	Y	-
Mesua ferrea	-	Y	-
Morus alba	Y	-	-
Picea smithiana	-	-	Y
Pinus roxburghii	Y	-	-
Pinus wallichiana	Y	-	-
Populus ciliata	Y	-	-

Prosopis juliflora	Y	-	-
Pterocarpus marsupium	-	-	Y
Quercus incana	Y	-	-
Quercus dilatata	-	-	Y
Quercus glauca	-	-	Y
Schleichera oleosa	-	Y	-
Shorea robusta	Y	-	-
Tectona grandis	Y	-	-
Toona ciliata	-	-	Y
Taxus baccata	-	Y	-

- The process of colour change of leaves before falling called autumn tint.
- Plants uprooted by wind called wind throw and able to withstand strong wind called wind firm.
- Term Forest succession was coined by Thoreau in 1863.
- Aubreville proposed Mosaic theory of climax.
- Standard tree classification adapted in Indian Forestry for regular forest crops is as follow:

Tree class	Class symbol	Abbreviation
Dominant tree:	I	D
-Predominant	–	D_1
-Codominant	–	D_2
Dominated tree	II	d
Supressed tree	III	s
Dead and Moribund	IV	m
Diseased tree	V	k

- Stand density index method was developed by Reineki in 1933.

CHAPTER 13

Agroforestry

- **Agroforestry** is a collective name for a and use system and technology whereby woody perennials as deliberately used on the same land management unit a agriculture crops and/or animals in same form of spatial arrangement or temporal sequence. In an agroforestry system there are both ecological and economical interactions between the various components (ICRAF, 1982).
- Term agroforestry was coined in 1977. International Council for Research in Agroforestry (ICRAF) came into existence in 1977.
- The National Commission on Agriculture emphasised agroforestry education in the **seventh** five-year plan period (1985-90).
- First agroforestry seminar in India was organised at Imphal in 1985. While international seminar on **"Agroforestry for Rural Need"** was organised in **1987**.
- First of all Post-graduation education in agroforestry was started in **Dr. Y.S. Parmar** University of Horticulture & Forestry, Solan (H.P).
- Similarly, Gujarat Agricultural University has started a P.G programme in agroforestry at the Aspee College of Horticulture & Forestry from **June 1993**.
- Benefits from agroforestry is classified into three categories:
  a. Environmental benefits
  b. Economic benefits
  c. Social benefits
- Home garden is also called as **multitier system.**
- Production and protection are two fundamental attributes of all agroforestry systems.
- **Allelopathy** is the direct or indirect effects of one plant to another through the production of chemical inhibitors that are released into the environment. The chemicals that impose influence are known

as allelochemicals. These chemical mostly are produced by leaves. These chemicals are secondary plant metabolites.

- There are four ways in which allelochemicals escape from plants viz.
  1. **Volatilization-** terpenes are released from the leaves of pine species by this process.
  2. **Leaching**- living or dead leaves of many plants contain growth inhibitors which escapes by leaching process.
  3. **Exudation**- roots of several crops and non-crops species release large quantities of organic compounds that inhibit the growth of other plant, these escape by exudation process.
  4. **Decomposition**- allelochemicals are released from plants residue by decomposition process.
- The species interaction due to chemical influences is also called as *Allelochemistry,* **Phytochemical ecology** or **Ecological biochemistry** and **Allelobiology**.
- Allelopathy is the best example for amensalism in plant.
- The process of taking nutrients from deeper soil profile and depositing them on the surface layer is referred to **nutrient pumping**.
- The process by which complex organic substances are broken down to simple inorganic elements is called as **mineralization.**
- The practices of rotational cropping helps in improving and maintaining soil fertility as well as soil and water conservation.
- The process of cycling of nutrients from soil to the plants and back to the soil is called as **nutrient cycling.**
- India has been classified into eight broad agroecological regions. These are given below:

Name	Annual precipitation	Soil groups	Remarks
1. Humid Western Himalayan	8-350 cm	Mountain meadow soils, Submontane meadow soils, Brown hill soils	Overgrazing, soil erosion and acidic soil
2. Humid Bengal-Assam Basin	220-400 cm	Alluvial red, Brown hill, Saline & Alkaline soils	Tropical dry deciduous forest & agriculture
3. Humid Eastern Himalayan Region And Bay Islands	200-400 cm	Brown Hill, Red loamy, Red & yellow alluvial and laterite soils	Shifting cultivation, Bamboo cultivation, Horticulture

4. Sub Humid Sutlej-Ganga Alluvial Plains	30-200 cm	Calcareous Srozem, Alluvial soil, Saline & Alkaline soils	Sal, Teak & bamboo vegetation
5. Sub Humid to Humid Eastern & South-Eastern Uplands	180-190 cm	Mixed soil groups	Sal, Teak plantation is dominant
6. Arid Western Plains	10-65 cm	Alluvial soils Black desert soils, Saline and Alkaline soils	Devoid of tree. Khejri is main crops
7. Semi-Arid Lava Plateau And Central Highlands	70-125 cm	Alluvial black soils, Lateritic red & black soils	Semi-evergreen forests (West. Ghat) Deciduous types
8. Humid To Semi-Arid Western Ghats And Karnataka Plateau	60-300 cm	–	Black wattles Blue gum species.

- According to **Nair (1987)** agroforestry system (**AFS**) can be classified into following sets of criteria:
  1. **Structural basis**-This considers the compositions of the components.
  2. **Functional basis**-This is based on the major function or role of system.
  3. **Socioeconomic basis**-This considers the level of inputs of management.
  4. **Ecological basis-**This is based on environmental conditions.
- Diagnosis & Design (D & D) is simply a systematic approach to the application of this principle in agroforestry. Three key features of D & D are **flexibility, speed and repetition**.
- The procedures of agroforestry D & D are usually done by two ways viz: *Macro D & D* and *micro D & D*. Macro D & D covers entire ecological zone within a country while micro D & D covers chosen ecological zone within a state.
- A good agroforestry design should fulfil the following criteria viz:

  1. Productivity 2. Sustainability 3. Adoptability
- Structural basis AFS can be classified into two categories viz:

  A. Nature of components

  B. Arrangement of components

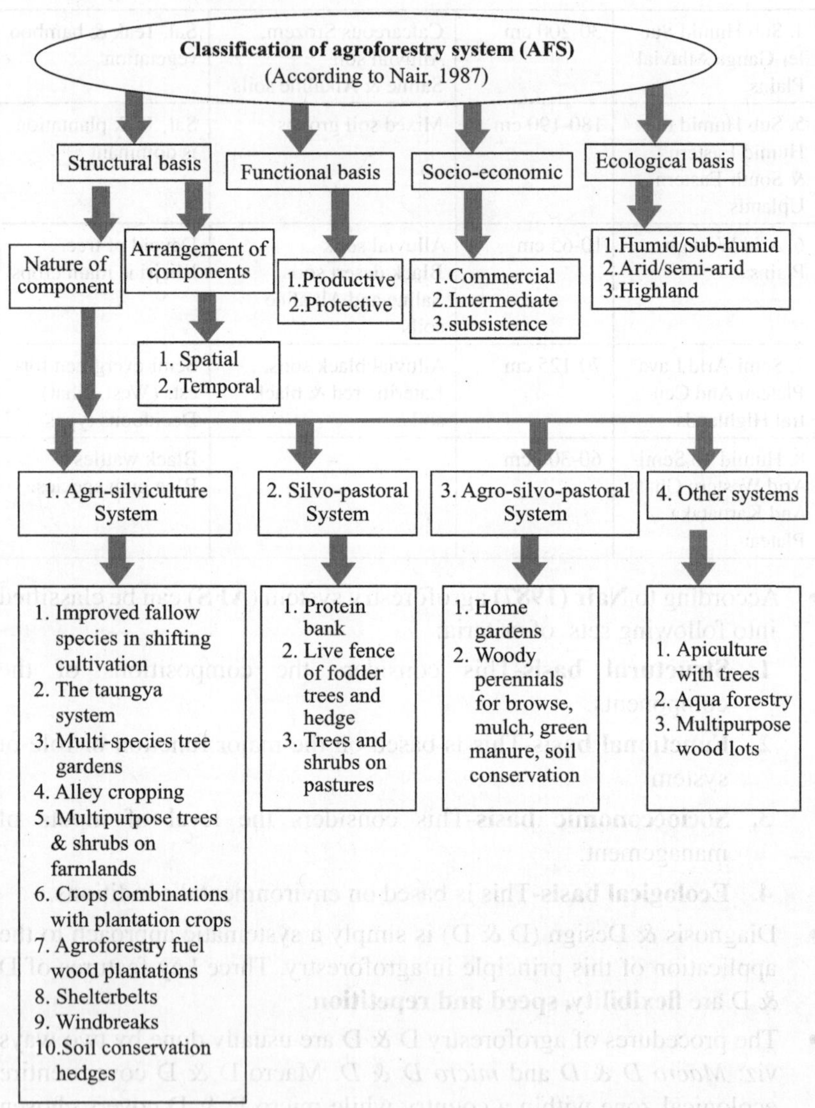

- **Improved fallow species in shifting cultivation:** Fallows are crops-land left without crops for periods ranging from one season to several years. The main objective is to recover depleted soil nutrients.
- **Multi-species Tree Gardens (MTG)** - The major function of this system is production of food, fodder and wood products for home consumption and sale for cash.

- **Alley cropping or Hedgerow Intercropping-** A method of planting in which rows of trees are interspersed with rows of crops, improving the soil and providing nutrients, particularly nitrogen, to the crops. Ideally hedgerows should be positioned in an east-west direction with 4 to 8 meter spacing between rows and 0.25 to 2 meter spacing between trees within rows. *Cassia siamea, Leucaena leucocephala, Gliricidia sepium, Calliandra calothyrsus* and *Sesbania sesban* are commonly used **legumes** tree species for alley cropping.
- **Shelterbelts-** Shelter belt is defined as a belt/blocks of trees or shrubs established at right angles to the prevailing wind. Ideal width of shelterbelts upto 50 meter. The ratio of height and width should be 1:10. A typical shelter belt has a triangular shape. The minimum length of shelterbelts should be about 25 times its height. Shelterbelts may be raised in quadrangles if wind direction tends to change very often.
- **Wind-break-** Wind-breaks are the strips of the trees and/or shrub planted to protect fields, homes, canals or other areas from wind and blowing soil or sand. The wind-breaks reduce the wind velocity between 25 to 75 per cent of the open wind speed. Wind-breaks should be planted at right angles to the wind from which protection is needed. Wind-breaks planted north-south gives better protection while planted in east-west gives better shading to crops. A 20 m tall wind-break provides protection from 100 m on the upwind side and 500 m on the downwind side. The minimum length of wind-breaks should be about 12 times the mature height of the trees. Wind-breaks of 3 to 5 rows are most effective. Species such as *Eucalyptus* and *Casuarina* will form effective winds breaks.
- The production of woody plants combined with pasture is referred as a silvopastural (trees+pastures) system. This is mostly practiced in dry areas.
- The production of agricultural crops along with woody plants & pasture is referred as a agrosilvopastoral (agricultural crops+ trees+pastures) system.
- **Home Gardens-** A home garden can be defined as a farming system that combines physical, social and economic functions on the area of land around the family home. This is one of the oldest agroforestry practices found extensively in high rainfall areas in tropical south and south-east Asia. Most commonly practiced in Kerala & Tamil Nadu with humid tropical climates and where coconut is the main crop.
- Fodder and legumes are widely grown to meet the daily fodder requirements of cattle in home gardens. In India every homestead has around 0.2 to 0.5 ha land for personal production.
- Home garden has usually 3-4 vertical canopy strata. Lowermost layer

(1.0 m height) is dominated by vegetables, intermediate layer (5-10 m height) have fruit plants while uppermost layer (10-25 m height) have full-grown timber trees.

- Most common agroforestry system in arid region is agro-silvicultural system. Agro-silvicultural (forestry+agronomy) is commonly practiced in both lowlands and uplands.
- Freshly cut wood contains about 23-25% moisture. The removal of a kg of water from wood involves an expenditure of 620-670 kcal.
- Coppicing and pollarding are forms of **heading back** carried out on the main trunk (stem).
- Closed nutrient cycles are known to operate in mixed evergreen natural forest.
- Land use factor is = C+F/C (where C- cultivation phase, F-fellow phase).
- Waldfeldban is like as taungya system in Germany.
- Ultimate object of taungya system is wood production meanwhile food production is ultimate object of shifting cultivation.
- Cut and carry fodder system called protein bank.
- Non leguminous N-fixing tree is Alnus nepalensis.
- **SALT-1**
- Sloping Agricultural Land Technology
- Also called 01 ha model of Agroforestry
- Ratio of Agricultural component : Forestry components is 75:25.
- All SALT models are developed by MBRLC.
- **SALT-2**
- Simple Agro-Livestock Technology
- Also called half ha model of goat based Agroforestry
- Ratio of Agricultural component : Livestock: Forestry components is 40:40:20.
- **SALT-3**
- Sustainable Agroforest Land Technology
- Also called 02 ha model of small scale reforestation
- Ratio of Agricultural component : Forestry components is 60:40.
- **SALT-4**
- Small Agro-Fruit Livelihood Technology
- Also called one half ha land model
- Father of Indian Agroforestry is B.S. Chundawat.
- Term taungya was coined by Upan Hle.
- Allelopathy term was coined by Hans Molisch.
- Macro and micro D&D takes 3 months to complete

CHAPTER 14

Social Forestry

- The word Social forestry was coined by **Westoby** and was used in the Ninth Commonwealth Forestry Congress in 1968.
- **China** was one of the first country to embark on a major community reforestation programme.
- **Social forestry** was first adopted successfully in the Gujarat state.
- According to **Prasad (1985)** "Forestry outside the conventional forests which primarily aims at providing continuous flow of goods and services for the benefit of people".
- Afforestation in the post-independence period can be divided in to three phases. In the first phase **'Van Mahotsav'** was started in fifties which failed to attract attention largely due to ignorance at all levels. In the second phase, farm forestry was started in some states in the 1970 s.
- The objectives of Social forestry adopted by the **Commission (1976)** were based on the economic needs of the community aimed at improving the conditions of living.
- The scope or components of social forestry defined by the Commission includes **farm forestry, extension forestry, reforestation in degraded forests and recreation forestry**.
- **Farm forestry**- Farm forestry was defined by NCA (1976) as the practice of forestry in all its aspects in and the around the farms or village lands integrated with other farm operations.
- **Extension forestry**- Extension forestry is the practice of forestry in areas devoid of tree growth and other vegetation situated in places away from the conventional forest areas with the object of increasing the area under tree growth.
- **Mixed forestry-** Mixed forestry is the practice of forestry for raising fodder grass with scattered fodder trees, fruit trees and fuel wood trees on suitable wastelands, panchayat lands and village commons.

- **Recreation forestry-** Recreation forestry is the practice of forestry with the object of raising flowering trees and shrubs mainly to serve as recreation forests for the urban and rural population. This type of forestry is also known as Aesthetic forestry which is defined as the practice of forestry with the objective of developing or maintaining a forest of high scenic value.
- Many tribal communities practice slash and burn agriculture in tropical and subtropical regions of Asia, Africa and Oceania. This consists of cutting down trees and setting them on fire and raising crops on the resulting ash called ***Jhuming cultivation or shifting cultivation or slash and burn cultivation.***
- Shifting cultivation in Assam, Tripura, and Arunachal Pradesh known as *jhum*, while in Odisha as *podu* or *dhai*, in M.P as *penda*, in Tamil Nadu as *podu* or *kumri*, in Kerala as *poonam*.

State name	Jhuming cycle (in years)
Arunachal Pradesh	1-17
Meghalaya, Mizoram & Tripura	4-5
Assam hills	5-10
Manipur	6-8
Nagaland	6-15

- Bewar, dhya, dippa, erka, jhum, kumri, penda, pothu, podu are other name of **shifting cultivation** in India. While in other countries it is known as Jumar (Java), Ray (Vietnam), Tam-ray (Thailand), Hanumo (Philippines), Chena (Sri Lanka), Karen (Japan, Korea) and Taungya (Burma-Myanmar).
- Taungya system is also called as **Shamba** in (East Africa), kumri, podu (Tamil), Jhum (Assam) and Ponam (Malayalam).
- Taungya is now applied to the method of raising forest plantations in combination with field crops and also known as **Agri-silviculture method**.
- The Taungyas can be of the following three types:

Departmental taungya	Leased taungya	Village taungya
Owner- Forest department	*Owner-* Lease person	*Owner-* Villagers
Duration- Not effective	*Duration-* 2 years	*Duration-* 2-5 years
Offered land- Not effective	*Offered land-* Auction need	*Offered land-* 0.8-1.6 ha.

CHAPTER 15

Genetics and Tree Improvement

Term	Given by
Genetics	W. Bateson
One mutant gene – one metabolic block	A. Garrod
Allele	W. Bateson
Factor	W. Johannsen
Element or Factor	G.J. Mendel
Phenotype and Genotype	W. Johannsen
Homozygous and Heterozygous	W. Bateson
Incomplete dominance	C. Correns
Lethal gene in plant	E. Baur
Lethal gene in mice	L. Cuenot
Polygenic inheritance	Nilsson-Ehle
Theory of Linkage	T.H. Morgan
Chromosomal theory of inheritance	Walter Sutton & Theodor Boveri
Coupling and repulsion theory	W. Bateson & R.C. Punnett
Sex chromosome	Mc. Clung
Chromosome	H.W.G. Waldeyer
X-chromosome (called x-body)	H.P. Henking
Allosome & Heterosome	Thas. H. Montgomery
Chromosomal theory of sex determination	H.B. Wilson & N.M. Stevens
Genic balance theory	C.B. Bridges
Eugenics	Sir Francis Galton (father of eugenics)
One gene one enzyme hypothesis	G.W. Beadle & E.L. Tatum

Jumping gene (in maize plant)	Mc. Clintock
Operon concept	F. Jacob & J.L. Monad
DNA finger printing/DNA test	Alec Jeffery
DNA finger printing/DNA test (in India)	Dr. Lal Ji Singh & Dr. V.K Kashyap
Cell lineage theory	Rudolf Virchow
Amitosis	R. Remak
Mitosis	W. Flemming
Meiosis	J.B. Farmer & J.E.S. Moore
Protoplasm	J.E. Purkinje
Lipid	Bloor
Nucleic acid term	R. Altman
Nucleic acid discovered by	F. Meischer
DNA term	E. Zacharis
DNA discovered by	F. Meischer
Double helix model (DNA)	J.D. Watson & F.M. Crick
Genetic code term	George Gamow
Genetic code discovered by	M. Nirenberg, H.J. Mathhai & H.G. Khorana
Central Dogma	F.H.C. Crick
Fluid mosaic model	S.J. Singer & G.L. Nicolson
Cytoplasm	E.A. Strasburger
Mitochondria term	C. Benda
Nucleus	Robert Brown
Cell	Robert Hook

- **Genetics–** collective study of heredity and variations.
- **Heredity–** transmission of genetic characters from parents to offsprings.
- **Variation**– individuals of same species having some differences.
- **G.J. Mendel**– father of genetics.
- **W. Bateson–** father of modern genetics.
- **Factor**– unit of heredity which is responsible for inheritance and appearance of characters.
- **Allele**– alternative form of a gene which are located on same position on the homologous chromosome.

- **Phenotype**– It is the external and morphological appearance of an organism for a particular character.
- **Genotype**– The genetic constitute or genetic make-up of an organism for a particular character.
- **Hybrid vigour/Heterosis**– superiority of offspring over its parents.
- Loss of hybrid vigour due to inbreeding is called as **inbreeding depression.**
- **Gregor Johann Mendel** was born on July 22, 1822 in Austria and died in 1884. He started his experiment on pea *(Pisum sativum)* in 1856.
- After 16 years of Mendel's death Carl Correns (Germany), Hugo de Varies (Holland) and Tschemark (Austria) rediscovered Mendelism.
- Mendel studied 7 characters (contrasting traits) which were present on four different pairs of chromosomes (1st, 4th, 5th and 7th).
- Gene which controls more than one character is called **pleiotropic gene.**
- Removal of stamen from the bisexual flowers at a juvenile stage is called **emasculation.**
- Checker Board method was first used by **Reginald. C. Punnet** in 1875-1967.

Cross name	Studied character	Phenotypic ratio	Genotypic ratio
Monohybrid cross	One trait	3:1	1:2:1
Dihybrid cross	Two pairs	9:3:3:1	1:2:2:4:1:2:1:2:1
Monohybrid Test cross	-	1:1	1:1
Dihybrid test cross	-	1:1:1:1	1:1:1:1

- A **back cross** is a cross in which F_1 individuals are crossed with any (either recessive or dominant) of their parents.
- When F_1 individuals is crossed with dominant parents then it is called **out cross.**
- When F_1 progeny is crossed with recessive parents then it is called test cross.
- Test cross helps to find out the genotype of dominant individuals.
- When two parents are used in two experiments in such a way that in one experiment "A" is used as male parent and "B" is used as female parent, while in other experiment vice-versa. Such type of a set of experiments is called **Reciprocal cross.**

- Allelic interaction takes place between allele of same gene which are present at same locus on gene.
- More than two alternative forms (due to mutation) of same gene are called as **multiple allele**.
- Gene which causes death of an individual when it comes in homozygous condition is called **lethal gene**.
- When a gene prevents the expression of another non-allelic gene then it is known as epistatic gene and this phenomena is called as **epistasis**.
- Collective inheritance of character is called linkage. Linkage was first observed by **Bateson & Punnett** in *Lathyrus odoratus.* Sex linkage was first discovered by **Morgan** in Drosophila & coined the term linkage.
- When the genes are present on sex-chromosome it is termed as sex linked gene and the phenomena is called as **sex-linkage**. Haemophilia, colour blindness are the example of x-linked gene.
- Haemophilia is also called **bleeder's disease** and was first discovered by **John Otto (1803)**. The gene of haemophilia is recessive and x-linked lethal gene.
- Genotype of carrier female is **($X^h X$)**, affected female is **($X^h X^h$)** and affected male is **($X^h Y$)**. While in colour blindness carrier female **($X^c X$)**, affected female **($X^c X^c$)** and affected male **($X^c Y$)**.
- XY female and **XX male** or **ZW female** and **ZZ male type** of sex determination is found in insects, birds, fishes and reptiles.
- **Eugenics** is a Greek word which mean **"well born"**. It mean improvement in human species through change in hereditary characters in a scientific manner.
- Sum total of genes in a reproductive gametes of a population is called **gene pool**.
- Migration of gene from one population to another population by cross fertilization is called **gene flow**.
- The existence within the population of disadvantageous allele in a heterozygous genotype is known as **genetic load**.
- **Gene frequency** is defined as proportion of different alleles of a gene in a population.
- **Hardy Weinberg** (1908) independently discovered that an equilibrium is established between frequencies of alleles in random mating population and these gene frequency remains constant from generation to generation.

$$p^2 + 2pq + q^2 = 1$$

- Segment of DNA which moves from one chromosome to another chromosome within the genome of an individual is called **jumping gene**.
- **An operon** is a part of genetic material or DNA, which acts as single regulated unit having one or more structural genes- an operator gene, a promotor gene, a regulatory gene.
- Dolly sheep was produced by using nuclear transfer techniques by **Dr. Ian Wilmut** at Roslin Institute of Scotland in 1997.
- Total gene component inherited haploid sets of chromosomes in an individual is called as **genome**.
- *Arabidopsis thaliana* was the first plant in which complete genome was sequenced.
- A *gene bank* is a store house of clones of known DNA fragments, genes, gene maps, seeds, spores, frozen sperms or eggs or embryos.
- A *gene library* is a collection of many of the desired genes of DNA fragments maintained in clones of bacterial or some other cells.
- Mitosis produced genetically identical cells, which are similar to mother cell. It occurs in somatic cells of an organism.
- Replication of nuclear DNA and synthesis of histone protein takes place in **s-phase**. While synthesis of different types of RNA & proteins takes place in G_1-phase.
- Mitosis is divided into **Prophase, Metaphase, Anaphase and Telophase**.
- Prophase is the longest stage of mitosis, where chromosomes become condensed and at the end of prophase cell do not show golgi complexes, ER, nucleolus and nuclear membrane.
- Morphological studies of chromosomes is most easily done in metaphase stage of mitosis.
- Anaphase is the **shortest** stage of mitosis. Division of centromere occurs in this stage.
- Mustard gas and Ribonucleases are **mitotic poisons**.
- Meiosis ensures the production of four haploid cells which are dissimilar to parent cell. In meiosis, nucleus divides twice and division of chromosome occur only once.
- **Prophase**-I of meiosis is divided into leptotene, zygotene, pacyhtene, diplotene and diakinesis. Pairs of homologous chromosomes are called Bivalents.

Sub-stage of prophase	Characteristics
Leptotene	Bouquet stage, longest & thinest chromosomes
Zygotene	Synapsis, synaptonemal complex
Pacyhtene	Tetrad conditions
Diplotene	Chiasmata visible (crossing over)
Diakinesis	Terminalisation of chiasmata

- Compound of protoplasm:

Inorganic compounds	Organic compounds
Water - 70-90%	Proteins - 7-14%
Salts, acids, bases, gases - 1-3%	Lipids - 1-3%
	Carbohydrates - 1-2%
	Nucleic acids, enzymes & other - 1-3%

- Hardest material in animal kingdom is **enamel** while in plant kingdom hardest material is **sporopollenin** (exine layer).
- First respiratory substance is **carbohydrates** [$C_x(H_2O)_y$]. Simple carbohydrates which are soluble in water and sweet in taste are called **sugars**.
- Ribose, ribulose, arabinose, xylulose and deoxyribose are the example of **pentose (monosaccharides) sugar**.

Name of carbohydrates	Known as
Glucose	Dextrose, Grape sugar, Blood sugar
Fructose	Fruit sugar
Galactose	Brain sugar
Maltose	Malt sugar (germinating seeds)
Sucrose	Invert sugar, Cane sugar, Table sugar
Glycogen	Animal starch
Pectin	Plant cement (plant middle lamella)

- **Lipid** do not form polymers. **Proteins** are heteropolymer of amino acid having peptide bond.
- Out of 20 amino acids 10 amino acids are not synthesized in our body so these must be present in our diet. These are called **essential**

amino acids. These are threonine, valine, leucine, isoleucine, lysine, methionine, phenylalanine, tryptophan, arginine and histidine.

- 10 amino acids, are synthesized in our body so these are called **non-essential amino acids**.
- **Glycine** is the simplest and Tryptophan is complex amino acid. All amino acids are laevo-rotatory except Glycine (non-rotatory).
- **Quaternary structure** is the most stable structure of proteins. Examples are haemoglobin, insulin.
- **Elasticity in** wheat flour is due to Glutellin protein. Nucleic acids are the polymer of **nucleotides (N_2-base + Pentose sugar + Phosphate).**
- Cytocine (C), uracil (U) and thymine (T) are **pyrimidines base** while adenine (A) and guanine (G) are **purine bases**.
- **DNA** have deoxyribose pentose sugar and four N_2-bases A, T, G and C. Both polynucleotide chains of DNA are **complementary** and **antiparallel** to each other.
- Adenine binds to thymine by two hydrogen bonds and cytosine binds to guanine by three hydrogen bonds.
- Out of two strand of DNA only one strand participates in transcription, it is called **antisense strand or non-coding strand or template strand**. While other strand which do not participate in transcription is called **sense strand or coding strand**.
- Right handed (clockwise twisting) DNA is **B-DNA** (10 base pairs) while left handed (anticlockwise twisting) DNA is **z-DNA** (12 base pairs) discovered by **Rich**.
- DNA molecule is Dextrorotatory while RNA is Laevorotatory.
- **Oswald Avery, Colin Macleod and Maclyn Mccarty** proved the genetic material is DNA.
- Semi-conservative mode of DNA replication was first proposed by Watson & Crick and experimentally proved by **Meselson & Stahl**.
- DNA replication takes place in **5 -3** direction with the help of DNA polymerase.
- RNA has ribose sugar in place of deoxyribose in DNA and **uracil** in place of thymine. RNA is **single stranded** while DNA is double stranded.
- Non-genetic RNA is of three type viz.-r-RNA, t-RNA and m-RNA.

Features	r-RNA	t-RNA (soluble RNA)	m-RNA
Discoverer	Kuntze	Hogland, Zemecknike	Jacob & Monad
Stability	Most stable	Stable	Least stable
Ribosome	80s	4s	-
Function	Provide attachment site to t- RNA (larger site) & m-RNA	Carrier of amino acids.	New polypeptide chain formation

- Formation of RNA over DNA template with the help of RNA polymerase enzyme is called **transcription**. Segment of DNA involved in transcription is called cistron. Transcription proceeds in $5 \rightarrow 3$ direction.
- The relationship between the sequence of amino acids in a polypeptide chain and nucleotide sequence of DNA or m-RNA is called **genetic code (64 codons)**.
- UAA (Ochre), UAG (Amber) and UGA (Opal) are non-sense codons or stop codons. AUG and GUG are **initiation codons.**
- The formation of m-RNA from DNA and then synthesis of protein from it, is known as **central dogma**. It means, it includes transcription and translation.
- Cell theory was proposed by **Schleiden** (Botanist) and **Schwann** (Zoologist).
- Cell-membrane is **selective-permeable** and living boundary of all living cells.
- **Endoplasmic Reticulum** (ER) helps in protein synthesis. **Rough ER** helps in protein synthesis while smooth ER helps in lipid synthesis.
- **Golgi body** termed as "*Director of macromolecular traffic in cell*" or *middle men of cell.*
- **Lysosomes** are filled with digestive enzyme termed as acid hydrolases ($P^H = 5$).it is also called as suicidal bags of cell.
- **Mitochondria** is called power house of the cell. Mitochondria and chloroplast are semi-autonomous cell organelle. **Oxysomes** or $F_1 - F_0$ particles are found on inner membrane of mitochondria.
- Red colour of chilli and tomato is due to red pigment called **lycopene.**
- **Amyloplasts** store carbohydrates, while elaioplasts store fats and oils and aleuroplasts store protein. Leucoplasts store nutrients. Sphaerosomes occur only in plants cell and store lipids.

- **Ribosomes** are made up of proteins plus nucleic acids (RNA) and are non-surrounded membrane. Ribosomes are also called *protein factory of cell.*
- **Nucleus** is considered as controller or director of cell.
- Ribosome, Basal bodies and Centriole are membrane less cell organelles.
- **Bruce Zobel** is called father of tree improvement.
- The trees produced from the seed of a parent tree are called its **progeny**.
- Group of populations that generally interbreed with one another and that intergrade more or less continuously are referred to as **race**.
- Individuals that are more closely related to each other than to other individuals in a population are called a **family**.
- A group of individuals within a family are referred to as **siblings**.
- The group of related individuals when only one parent is common is called as **half-sib family**, when both parents are common called **full sib family**. An open-pollinated family is one in which one parent is common and other parent (s) are unknown.
- Variation in a population is **increased** by mutation & gene flow and **reduced** by natural selection & genetic drift.
- An **ecotype** is a group of plants of similar genotype that occupy a specific ecological niche. The concept of ecotype was suggested by Turreson (1922) who defined it as *genotypical response of a species to a particular habitat.*
- Cline was first defined by **Huxley (1938)** as *a gradient measurable characteristic.*
- A **land race** is a population of individuals that has become adapted to a specific environment in which it has been planted.
- **General combining ability** (GCA) is defined as the average performance of the progeny of an individual when it is mated to a number of the other individuals in the population.
- **Specific combining ability** (SCA) is defined as the average performance of the progeny of a cross between two specific parents that are different from what would be expected on the basis of their general combining abilities alone. It can be negative or positive.
- The breeding value of an individual is defined as twice its **general combining ability** (GCA).

- **Heritability** is a ratio indicating the degree to which parents pass their characteristics to their offspring. It is of two types viz. narrow sense and broad sense heritability.
- **Broad-sense heritability (H^2)** is defined as the ratio of total genetic variation (V_G) in a population to the phenotypic variation (V_P). Broad-sense heritability can range from 0 to 1.

$$H^2 = V_G / V_P$$

- **Narrow-sense heritability (h^2)** is defined as the ratio of additive genetic variation (V_A) in a population to total variation (V_P). Narrow-sense heritability can range from 0 to 1.

$$h^2 = V_A / V_P$$

- Selection differential is the difference between the ***mean*** of the selected individuals (*Xs*) and the population mean (*X*).

$$S = Xs - X$$

- **Genetic gain (*G*)** is the multiply product of narrow-sense heritability (h^2) and selection differential (*S*).

$$G = h^2.S$$

- **Mass selection** involves choosing individuals solely on the basis of their phenotypes, without regards to any information about performance of offsprings or siblings.
- Family selection involves the selection of entire family on the basis of their average phenotypic values.
- **Sib selection** involves selection of individuals on the basis of the performance of their siblings not on their own performance.
- **Progeny testing** involves the selection of parent trees on the basis of performance of their progeny.
- Selection of individuals on the basis of their deviation from the family mean, and family values per se are given no weight when selection are made is called **within family selection.**
- **Family within family** selection is two stage method involving selection of families followed by selection of individuals within families.
- Tandem selection, independent culling, selection index and recurrent selection are the methods of **multiple traits selection.**
- The selection procedure that involves many cycles of selection and breeding is known as **recurrent selection**.

- A tree that has been selected for grading because of its desirable phenotypic qualities but that has not yet been graded or tested is called **candidate tree**.
- Morphologically superior tree is called plus tree. While morphologically as well as genetically superior tree is called **elite tree**.
- Trees that are located in the same stand having nearly same age and growing on the same or better site as the select tree and against which the select tree is graded is called **comparison or check tree**.

CHAPTER 16

Tree Seed Technology

- **Seed development** and **maturation** of tree seeds are greatly affected by external factors like birds, insects etc.
- Seed - Fertilized mature ovule is called **seed**. Seed is a link between two generation.
- **Cleaning** and **grading** can improve physical quality of a seedlot.
- Sal, bamboo and teak fruits are considered as seed. Teak seed are non-endospermic.
- **Seed maturation** – Morphological, physiological and functional changes that occurs from the time of fertilization until the matured ovules (seeds) are ready for harvest. (MC=15 to 20 %).
- **Maturity** is defined as the point of maximum dry weight of seed. It occurs when seeds have high moisture content 35-40 %.
- **Harvestable maturity** – seed first time dries to a level of 14 % or less
- Tectona grandis seed mature in **120 – 200** days.
- Common seed colour is **brown** and brown derivatives in forest trees.
- Winged seeds are found in *Dipterocarpus, Terminalia, Acer* and *Pterocarpus.* Seeds of most conifers are winged.
- **Hairy seeds** are found in Bombax, Salix, and Populus. While seeds of Apocynaceae family are hairy.
- **Hard seed** coat are the characteristics of legumes trees.
- Indian Seed Act was formed in **1966**. It has 25 sections. While Seed Rules was formed in **1968**. Seed order was formed in **1983**.

Seed name	Produced by	Purity	Certification	Colour (Tag)
Nucleus seed	Nucleus seed	G & P-100%	No need	-
Breeder seed	Nucleus seed	G & P-100%	No need	Yellow
Foundation seed	Breeder seed	G-100% & P-98%	Required	White
Registered seed	Foundation seed	G-100% & P-98%	Required	Purple
Certified seed	Registered seed	G-100% & P-98%	Required	Azar blue

- Systematic removal of individuals not desired for the perpetuation of the population is called **roughing**.
- **Imbibition** is the first step during seed germination. In 1968, Amen presented the first model of the sequence of biochemical events in seed germination.
- Maximum seed germination occurs in **red light** and minimum in **far-red light**.
- Branches developed after removal of surrounding trees are called **epicormics branches**.
- Table showing differences between CSO and SSO:

Characteristics	Seedling Seed Orchard (SSO)	Clonal Seed Orchard (CSO)
Origin	Seedlings	Vegetative propagation
Genetic base	Broader	Narrower
Selection differential	Less	Higher
Testing	On family level	On individual level
Trees used for testing	100 or more parental trees	20-40 clones
Outstanding genotype	Appear only once	Many types

- Seeds which has life span of more than 15 days are called **orthodox seeds**.
- Those seed which cannot be dried to moisture contents below 30 per cent without injury and are unable to tolerate freezing are called **recalcitrant seed**.
- Seed can be categorized into three biological classes according to their period of storability:

Seed class	Life span of seeds in Year
Microbiotic	Upto 3 years
Mesobiotic	3-15 years
Macrobiotic	>15 years

Orthodox seeds	Recalcitrant seeds
Acacia, Prosopis, Albizia, Cassia, bauhinia, Pinus, Picea, Pongamia, Pterocarpus, Bambusa arundinacea and eucalyptus etc.	*Quercus, Syzygium, Swietenia, Azadirachta and Aesculus indica etc.*

- There are a number of forces that alter the variation pattern within a population. **Variation** is increased by mutation and gene flow. While reduced by natural selection and genetic drift in the natural environment.
- Mostly vegetative orchards have been established by **grafting**.
- First generation orchard usually results from selection from the **natural stands** or **unimproved plantations**.
- **1.5 generation orchards** have best genotypes from a number of orchards of similar geographical background and bringing them together into a new greatly improved generation orchard.
- 1.5 generation orchards are also known as **improved first generation orchard**.

CHAPTER 17

Soil Science

- **Soil** is a dynamic body developed by the natural forces acting on natural materials. It is usually differentiated into horizons from minerals and organic constituents of variables depth which differ from the parent material below in morphology, physical properties and constituents, chemical properties and composition and biological characteristics.
- Soil has four components viz. two are solid components and two are liquid components.

Soil components	Nature	Composition per cent
Mineral matters	Solid	45
Organic matters	Solid	05
Water	Liquid	25
Air	Liquid	25

- A vertical section of soil shows five master **horizons** (O, A, B, C and E/R).

Horizon name	Other names
O	Organic horizon (present in forest soil)
A	Washing out or Elluvial horizon or leaching horizon
B	Washing in horizon or Illuvial horizon
C	Unconsolited materials
E/R	Unweathered rocks

- Earth shows three layer in their internal structure:

Earth layer	Thickness (in km)	Density (gm/cm^3)
Crust	5-56	2.6 - 3.0
Mantle	2900	3.0 - 4.5
Core	3500	9.0 - 12.0

- Density of earth is **5.5 gm/c.c** while density of rock is **2.6-2.7 gm/c.c**.
- Mixture of two or more minerals called **rocks**. Study of rocks called **petrology**.
- Study of origin, classification and description of soil is called **pedology**.
- Vertical section through a soil is called **soil profile**.
- A+B horizons collectively called as **solum**.
- **Hydrolysis** is the most important chemical weathering process.
- The relative proportion of sand, silt and clay in the soil know as soil **texture**.
- Mass of unit volume of soil including the pore space is called as **bulk or apparent density**.
- CO_2 content of soil is **0.3%**. It is ten times more as compare to atmospheric CO_2 (0.03%).
- Soil colour chart is given by **Munsell**. It has hue, chroma and value variables.

Colour variables	Shows
Hue	Dominant spectral colour
Chroma	Purity of colour
Value	Lightness or darkness of colour

- Soil water retained between **-0.33 bar & -15 bars** is called available water. While capillary water is retained at **-1/3 to -15 bar**.
- Movement of water into the soil is called as **infiltration**.
- Amount of water required to produce a unit quantity of dry weight material is called as **water use efficiency**.
- Capacity of the soil to change its shape under moist conditions is called as **plasticity**.
- **Molybdenum** is not found in acidic soil.
- The power to resist a change in pH by the soil is called its **buffering capacity**.
- The process of removal of silica instead of sesquioxides to concentrate in the solum is called as **laterisation**.
- **Saline soils** also called as solon chalk or white alkali soils.
- **Alkali soils** also called as sodic soils or solonetz or non-saline alkali soils.

- Particle density or absolute density or gain density and bulk density are expressed in term of **mg/m^3 or gms/cm^3**.
- The weight of per unit volume of the solid portion of soil is called **particle or real density**.
- The mass (weight) per unit volume of dry soil is called as **apparent or bulk density**.
- Particle density (2.65 gm/c.c) is always higher than bulk density (1.33-1.65 gm/c.c).
- Number of land suitability classes is **2**.
- Total number of soil irrigability classes are **5**.
- **Poudrette** is called as night soil manure.
- **Gypsum** is used for reclamation of alkali soils.
- Sandy loam soil is light textured. Specific gravity of soil is **0.2** and for water it is **1.0**.
- One gram of soil contains **14 millions** of microorganisms.
- The **macrospores** in the soil generally constitute of soil air.
- **Vertisols** is related to black soils. Red soil is found in Tamil Nadu.
- **Fossils** are found in sedimentary rocks.
- The length of engineer chain is **100 ft**. while Gunter chain is 66 ft.
- **Red soils** are mainly falls under the Alfisols.
- **Clay soils** can become as hard as a stone on drying.
- B dries out quickly after a rainstorm.
- The transfer of minerals from the top soil to sub soil through water is called as **leaching**.
- **Phosphorous fertilizer** are fixed immediately after application in the soil.
- *Aspergillus niger* is used to test the **potassium element** status in the soil.
- Cation exchange capacity is expressed in term of **me/100 g or c mol/ kg**.
- **Soil texture** is an unchangeable soil property.
- The arrangement of primary particles and their aggregates into a certain definite pattern is called as **soil structure**.

- Table showing retained potential of water at holding stage in sol:

Soil water	Retained potential at (in Bar unit)
Gravitational water	-1/3 bar
Capillary water	-1/3 - -31 bar
Hygroscopic water	>-31 bar
Available soil water	-1/3 - -15 bar
Unavailable soil water	>-15 bar

- Classification of soil silicates clay into 4 categories on the basis of Aluminium and silica sheet arrangement:

Silicate groups	Examples
1:1 types	Kaolinite
2:1 types	Montmorillonite
2:1 non-expanding types	Hydrous mica or illite
2:2 types	Chlorites

- Amount of exchangeable cations per unit weight of dry soil is called **Cation exchange capacity** (CEC). The unit of CEC is **C mol kg^{-1} soil**.
- Soil survey has been categorised in to four types. These are Detailed soil survey, Reconnaissance soil survey, Detailed survey-reconnaissance soil survey and Semi-detailed survey.
- Cadastral maps having scale **1:8,000 or 1:4,000** are used in Detailed soil survey while topographical maps having scale **1:50,000** are used in Reconnaissance soil survey.
- Capacity of a soil to absorb or release anion under normal soil conditions is called **Anion exchange capacity (AEC)**.
- Soil formation equation was first given by **Dokuchaiev (1889)** later it was explained by **Jenny (1941)**. Jenny equation is given below:

  $S = f(cl, b, r, p, t)$ where, *b*-biosphere, *cl*-climate, *r*-relief, *t*-time, *p*-parent material, *f*-function and *S*-soil formation.
- Table showing 12 soil orders proposed in the 7th Approximation and their derivatives:

Soil order names	Symbols	Sub-orders
Entisol	Ent	4
Vertisol	Ert	4
Inceptisol	Ept	5
Aridosol	Id	2
Mollisol	Oll	7
Spodosol	Od	4
Alfisol	Alf	5
Ultisol	Ult	5
Oxisol	Ox	5
Histosol	1st	4
Andisol	and	7
Gelisol	el	3

- The foliar spray of urea have **1.5 per cent** biuret concentration.
- The process of accumulation insoluble lime salts (mostly Ca & Mg) in soil horizons is called **calcification**.
- The ratio of the weight of organic carbon to the weight of the total nitrogen in soil or organic material is called **C : N ratio**.
- **E-value** is used for determination of available phosphorus in soils.
- **Flocculation** is a process in which colloidal particles tend to attract each other and coagulate like a bigger particle.
- The process of accumulation of iron (Fe) and aluminium (Al) oxides in soil horizons to form laterite soil is called **laterization**.
- **L-value** is used to calculate the phosphorus supply to the soil.
- Formation of inorganic substances from organic substances by the microbial decomposition is called **mineralization**.
- The downward movement of water through the soil is called **percolation**.
- The disintegrated and decomposed mass of weathered rocks and soil material on the earth surface in which the soil develops is called **regolith**.
- A-value was given by **Fried & Dean (1952)** and Y or L-value was given by **Larson (1952)** is used for nutrient amount determination in the soil.

- **Soil erosion** is defined as the detachment and transportation of soil mass from one place to another place through the action of wind, water and beating action of raindrops.
- **Raindrop splash erosion, sheet erosion, rill erosion** and **gully erosion** are the types of soil erosions by water.
- **Raindrop splash erosion** results from soil splash caused by the impact of falling rain drops.
- Removal of a fairly uniform layers of surface soil by the action of rainfall and run-off water called **sheet erosion**. Raindrop splash erosion is first stage of water erosion.
- Removal of surface soil by running water with the formation of narrow shallow channels that can be levelled or smoothed out completely by normal cultivation is called **rill erosion**.
- Removal of surface soil by running water with the formation of channels that cannot be smoothed out completely by normal agricultural operation or cultivation is called **gully erosion**. Gully erosion is an advanced stage of rill erosion.
- **Saltation, suspension** and **surface creep** are the types of soil erosions by wind.
- **Saltation** is a process of movement of soil particles having size ranging from 0.05 to **0.5 mm** in a series of bounces or jumps.
- **Surface creep** is the rolling or sliding of large soil particles along the ground surface.
- **Suspension** represents the floating of small sized particles in the air stream.

Wind erosion types	Particle size (in mm)	Soil loss (percent of wind erosion)
Saltation	0.1-0.5	50-70
Surface creep	> 0.5	5-25
Suspension	< 0.1	3-4

- The rate of soil erosion in India is **16.4 t/ha.** 'O' horizon is absent in Arable soils.

- Table showing types of rocks with their examples:

Rocks	Examples
Igneous	Granite and Basalt
Sedimentary	Limestone, Sand stone and Dolomite
Metamorphic	Gneisis, Marble, Quartz and Slate

- Classifications of soil particles size:

Classification	According to IISS diameter (in mm)
Gravel	>2.0
Coarse sand	0.2-2.0
Fine sand	0.02-0.2
Silt	0.02-0.002
Clay	<0.002

- **Clay soils** have highest pore space than other soils.
- Soils of India is broadly classified into eight groups. Four majors are given below:

Soil group	Formed from	Occupied area (in mha)	Remarks
Alluvial soils	Entisols	143	Largest soils groups
Black soils	Vertisol	55	2nd largest group, called ***Regur***
Red soils	Alfisols	15	Called early soils
Laterite soils	Ultisols	25	Maximum leached soils

- **Soil series** is the smallest unit of soil classifications.
- **Organic matter** is found in highest quantities in grass lands than forest area.
- Organic matter = organic carbon (OC) × 1.724 (***Bemlen factor***).
- Humus and organic matter contains nitrogen 5 to 5.5% while **contain** carbon is **50 to 58%**.
- Highest acidic soils are found in **West Bengal**.
- When erosion is caused by excessive grazing, deforestation is called **anthropogenic erosion**.
- **Gypsum** is used for reclamation of alkali or sodic soils.

- Soils having atleast 20% organic matter is called as **organic soils**. It is generally less than 0.5% in Indian soils.
- **Zero tillage practice** was first started in USA.
- Parameters of different problematic soils shown in table:

Parameters	Saline soils	Alkaline soils	Saline-Alkali
pH	< 8.5	> 8.5	> 8.5
EC (ds/m)	> 4	< 4	> 4
ESP	< 15	> 15	> 15

- Nutrient contents of some important manure is given bellow:

Products	N	P	K
FYM	0.5	0.2	0.5
Town compost	1.5	0.4	1.4
Vermicompost	3.0	1.0	1.5
Night soil	5.5	4.0	2.0
Ground nut cakes	7.3	1.5	1.3
Safflower cakes	7.9	2.2	1.9

- The C: N ratio of humus is **10:1** while for normal soil, it is **12:1**.
- **K** and **Na** is determined by flame photometer.
- Urea contain **46 %** N, thiourea-**36.8 %** N and Urea formaldehyde (UF)-**38-42 %** N.
- **SSP** contain - 16 % P_2O_5, 19 % Ca, 12 % S. **DAP**- 18 % N, 46 % P_2O_5. **MAP**- 11 % N, 48 % P_2O_5. Ammonium nitrate contain 33 % nitrogen.
- Calcium ammonium nitrate (CAN) is also known as **Kissan Khad** contain 26 % N.
- Table showing transported soils:

Transported soil by	Called as
Gravity	Colluvial
Water	Alluvial
Ice	Glacial
Wind	Eolian

Soils	Sand (%)	Clay (%) + Silt (%) or both
Sandy soils	85	15
Loamy soils	70	30
Silt soils	10	90 silt

- Land capability classification (LCC) has eight classes of soils:

Soil class	Suitability	Used for	Class colour
I.	Very good land	Cultivation	Light green
II.	Good land	Cultivation	Yellow
III.	Moderately good land	Cultivation	Red
IV.	Fairly good land	Restrict cultivation	Blue
V.	Not good land	Pasture and grassland	Dark green
VI.	Not good land	Pasture and grassland	Orange
VII.	Not good land	Pasture and grassland	Brown
VIII.	Not good land	Wildlife & watershed	Purple

- Class I, II & III are suitable for cultivation while and V, VI, VII & VIII are not suitable for agriculture or cultivation.

CHAPTER 18

Agricultural Extension

S.No.	Programme	Year
1.	Community Development Programme (CDP)	1952
2.	National Extension Service (NES)	1953
3.	Pachayati Raj	1959
4.	Intensive Agriculture District Programme (IADP)	1960
5.	Hill Area Development Programme (HADP)	1960
6.	Tribal Area Development Programme (TADP)	1962
7.	Intensive Agricultural Area Programme (IAAP)	1964
8.	High Yielding Variety Programme (HYVP)	1965
9.	Farmers Training Education	1966
10.	Small Farmers Development Agency (SFDA)	1969
11.	Marginal Farmers and Agricultural Laborers Development Agency (MFAC)	1969
12.	Drought Prone Area Programme (DPAP)	1970
13.	Rural Works Programme (RWP)	1971
14.	Pilot Project Tribal Development	1972
15.	Employment Guarantee Scheme (EGS)	1972
16.	Command Area Development Programme (CADP)	1973
17.	Minimum Needs Programme (MNP)	1974
18.	Antyodaya	1977
19.	Desert Development Programme	1977
20.	Integrated Rural Development Programme (IRDP)	1978
21.	National Rural Employment Programme (NREP)	1980

22.	Training of Rural Youth for Self-employment (TRYSEM)	1980
23.	Self-Employment to Educated Unemployed Youth Programme (SEEUYP)	1983
24.	Rural Landless Employment Guarantee Programme (RLEGP)	1983
25.	Jawahar Rozgar Yojana (JRY)	1984
26.	Indira Aawaas Yojana	1986
27.	Rajiv Gandhi National Drinking Water Mission	1991
28.	Jawahar Gram Samridhi Yojana	1999
29.	Swaran Jayanthi Gram Swarojgar Yojana	1999
30.	Employment Assurance Scheme	1993

- Agriculture extension was first started in **Bihar Agricultural College, Sabour**.
- Etawah pilot project was headed by **Albert Mayer** in 1948.
- T and V programme was introduced in **1974**.
- **Masanobu Fukuoka** is the father of organic farming.
- National Agricultural Science Museum is located in New Delhi, and was established in 2004 and inaugurated by **Dr. A.P.J. Abdul Kalam** on 3rd November.
- Year **2002** is known as the year of Technology.
- ICAR was established on the recommendation of Lord Linlithgow Commission.
- The first KVK was established at Ponduchery in **1974**.
- Lab to Land programme was started by ICAR in **1979** on Golden jubilee.
- Television broadcast for rural development in India was started in **1977**.
- Shantiniketan project was started by **R.N. Tagore** in West Bengal.
- *Grow more food* campaign was stared in **1948**.
- T and V system of extension was started by **D. Benor**.
- Sevagram project was started by **M.K. Gandhi**.

- National Academy of Agriculture Research Management is located in **Hyderabad**.
- Working model is also known as **Mock-up**.
- The Royal Commission came in **1928**.
- *Kissan Bharti* periodical is published from **Pantnagar (U.K)**.
- The full form of **ATMA** is Agriculture Technology Management Agency.
- Extension Education is both discipline and **profession**.
- The Government of India set up planning Commission during **1954**.
- **NATP** was funded from World Bank.
- Sompal was the first chairman of National Commission on farmers.
- **M.S. Swaminathan** is the present chairman of National Commission on Farmers.
- Operational Research Projects were initiate during **1970-1971**.
- *National Food for work* programme was launched on **14th November 2004**.
- Role of different agencies for village development was included in **Chapati diagram**.
- **NABARD** is the Apex bank providing rural credit in Agriculture field.
- **Village** is the functional unit of rural society.
- First Five year plan stared during was **1951-56**.
- First rural development programme started was **Srinikenatan project**.
- The principle of *one village one society* was suggested by **Maclogen committee**.
- Panchayati Raj was first started in **Rajasthan**.
- KVK was recommended by **Mohan Singh Mehta Committee**.
- The National Agricultural Insurance Scheme (NAIS) was stated in **1999-2000**.
- **NARS** refers to National Agricultural Research System.
- **IVLP** stands for Institutional Village Linkage Programme.
- National Institute of Agricultural Marketing (NIAM) is located in **Jaipur**.

- Reserve Bank of India was established on 1st **April, 1935**.
- Full form of **SIDBI** is Small Industries Development Bank of India.
- National Food Security Mission is presently implemented in **17 states**.
- **Varsha Bima** is a rainfall insurance scheme of AICIL.
- The community development programme was started in India on **2nd October, 1952**.
- 2010-2015 is the recommended period for adopting and implementing **13th Finance Commission**.
- The first fully Indian Bank was **Punjab National Bank**.
- The time period of 11th Five year Plan is **2007-2012**.
- **KVIC** (khadi and Village Industry Commission) was established in Second five year plan.
- Buffer stock is maintained by **Food Corporation of India (FCI)**.
- Co-operative credit societies act was passed in India during **1912**.
- National Policy for Farmers was approved during **2007**.
- The term Extension was first used in **U.K.**
- Goal of extension is to promote **scientific out-look**.
- Success of rural projects depends upon **Agricultural Extension**.
- Institute for Studies on Agricultural and Rural Development **(ISARB)**.
- Food for work programme was renamed as **NREP**.
- The first **Kshetriya Gramin Bank** was opened in India during 1975.
- The first state that adopted two tier panchayat raj was **Karnataka**.
- IADP program is also called **package programme**.
- **Indian Farming** is a journal published in the journal of ICAR.
- TRYSEM was introduced in the year **1979**.
- The chairman for national development of council in India is **Prime Minister**.
- NARP was launched by ICAR in **1979**.
- The Cartagene Kyoto protocol for agriculture product came in the year **2005**.
- The political personality who got Norman Borlaug award was **C. Subramanium**.

- Nobel Prize for Micro Credit was given to **Mohammad Yunus**.
- Directorate of plant protection, quarantine and storage is located at **Faridabad**.
- Sub stations of NBPGR are **5**.
- Marthandam project was started by **Spencer Hatch**.
- **State farming** is the system of farming in which farmers are managed by Govt. and the **agricultural workers** are paid wages generally on monthly basis.
- Model village concept was given by **Daniel Hamilton**.
- **CAMPA** (Compensatory Afforestation Fund Management and Planning Authority) - CAMPA is meant to promote afforestation activities in order to compensate for forest land diverted to non-forest uses.
- **Carbon Stock** is defined as the amount of carbon stored in the ecosystem of the forest especially in living biomass and soil.
- The important characteristics of the adopters categories are mentioned in brief:

Categories	Per cent	Characteristics
Innovators	2.5	First to adopt, very few in number
Early adopters	13.5	Localite and doesn't test untried ideas
Early majority	34	Adopt after ideas just before avg. members of community
Late majority	34	Adopt new ideas just after avg. members of community
Laggards	16	Traditional, belongs to backword classes

- The adoption of an innovation usually follows a normal bell shaped curve when plotted it **time vs. frequency**. If the cumulative number of adopters is plotted instead of frequency, then result is an **S-shaped curve** (sigmoid curve).

- Important projects and associated persons are given below:

Project name	Year	Associate person
Gurgaon	1920	F.L. Bryne
Shantiniketan	1921	R.N. Tagore
Marathondom	1921	Spencer Hatch
Sewagram	1929	M.K. Gandhi
Etawa pilot project	1948	Albert Mayer
Nilokheri experiment	-	S.K. Dey
Firka Development	1946	T. Prakashan
Majdoor Manjil	1947	S.K. Dey

- Leagnes is the father of extension. Extension is a **Latin** word.

CHAPTER 19

Agricultural Statistics and Economics

- Mean, median and mode are same in **normal distribution**.
- **Randomized Block Design** is the best design for field experiments.
- The square of the standard deviation is known as **Variance**.
- The relationship between independent and dependent variables is computed by **regression coefficient**.
- The **regression coefficient** is independent of change of origin.
- The value of **chi square (X^2)** is always positive which ranges from 0 to infinity.
- The correlation coefficient is the **geometric mean** between two regression coefficients.
- The range of multiple correlation coefficient lies between **–1 to +1**.
- **Completely Randomized Design** is the most appropriate for the laboratory experiments.
- Local control is not applied for the design of CRD.
- Normal distribution is due to the work of **Demoivre**.
- If any items of the series is zero, geometric mean of **central tendency** become zero.
- A measure of the peakedness or convexity of a curve is known as **kurtosis**.
- The allocation of the treatments to the different experimental units in a random manner is known as **Randomization**.
- **RBD** is appropriate when the fertility gradient of the field is in one direction.
- **LSD** is appropriate when the fertility gradient of the field is in two directions.
- **R.A. Fisher** is considered as father of statistics.

- **Poisson distribution** deals with discrete variables.
- The minimum error degree of freedom for RBD should be at least **12.**
- The hypothesis of no difference is known as **Null hypothesis**.
- The probability of committing type 1 error is known as **Level of significance**.
- **Z test** is used to test the significance of the difference between two means z-test.
- For comparison of two means from independent samples **t-test** is applicable.
- The values of regression coefficient lie between **–infinity to +infinity.**
- The type of distribution in which mean and variance are equal is called **normal distribution.**
- In hypothesis testing the most common level of significance that is used by statisticians is **0.05**.
- When two variables moves in same direction, correlation is said to be **positive**.
- t-test is applicable when the number of treatments are **2.**
- **Principles of Experimental design** was given by R.A. Fisher.
- **Anova** was first proposed by R.A. Fisher.
- The simplest measure of dispersion is **range**.
- When some of the plots data is missing the most suitable experimental design is **Missing plot technique**.
- The statistical test used to determine the goodness of fit is **chi square test.**
- **Unit free relative** measure of dispersion is coefficient of variation.
- The curve of normal distribution is **bell shaped**.
- The correlation coefficient will be **negative** when x increases and y decreases.
- Binomial distribution is called **Bernoulli's distribution**.
- Chi square test is **parametric** test.
- Error degree of freedom for RBD is **(n – 1) (r – 1)**.
- The Indian economy is a type of **mixed economy**.
- **Net capital ratio** is total assets/total liabilities.
- National Institute of Agricultural Marketing (NIAM) is located in **Jaipur**.

- RBI was established on **1 April, 1935**.
- Full form of **SIDBI** is Small Industries Development Bank of India.
- The most valid law in agriculture production is **Law of diminishing returns.**
- The formula for rate of turnover is total expenses/gross income.
- Balance sheets provides a picture called **net worth.**
- The operational date of TRIPS agreement of WTO is **1 January, 2005**.
- National Income in India is estimated by **Central Statistical Organization (CSO).**
- Headquarter of SIDBI is situated in **Lucknow, U.P.**
- The main security guard of International Trade is **WTO**.
- NABARD was established in **6 five year plan**.
- C. Rangarajan was the chairman of **12th Finance Commission.**
- Father of economics is **Adam Smith.**
- **VAT** is an Indirect tax.
- Table showing experimental design:

Experimental design name	Local control	Applied to	Error	Remarks
Completely randomised design (CRD)	Not used	Homog-enous materials	$(n-1)$	One way classification
Randomised block design	One way control	Upto 20 treatments	$(n-1) \times (r-1)$	two way classification, block = treatments
Latin square design	Used	5-8 or 5-12 treatments	$(n-1) \times (n-2)$	Rows = columns Replication = treatment
Split plot design	Used	Sub plots	-	-

- **ANOVA** (analysis of variance) concept was given by R. A. Fisher.
- Table showing test of significance:

Test name	Given by	Applied to
Chi-square	Karl Pearson	Sample size ≥ 50
z-test	R.A. Fisher	Sample size > 30
Student t-test	W.S. Gosset	Sample size < 30

- Rejecting null hypothesis (H_0) when it is true or accepting alternative hypothesis (H_1) when it is false called is **Type error (Alfa)-α.**

- Accepting H_0 when it is false or rejecting H_1 when it is true called is **Type error (Bita)-β**.
- **Degree of freedom** is the difference in total number of items (n) and linear constraints (k).
- **Goodness of test** is used for distribution of Chi-square test.
- Table showing range of important test or distribution:

Tests/Distribution	Range
t-test	$-\infty$ to $+\infty$
Normal distribution	$-\infty$ to $+\infty$
Poisson distribution	$0 \leq x \leq \infty$
f-test	$0 \leq x \leq \infty$
Correlation coefficient	–1 to +1
Regression coefficient	$-\infty$ to $+\infty$
Probability	0 to 1

CHAPTER 20

Key Points of State Forest Report 2015

- Total forest cover in India is **7,01,673 sq. km** (increase of 3775 sq. km).
- Total forest cover as percentage of geographical areas **21.34 per cent**.
- Total tree cover in India is **92,572 sq. km** (increase of 1306 sq. km).
- Total tree cover as percentage of geographical area is **2.82 per cent**.
- State having largest total forest cover in India is **Madhya Pradesh** having **77, 462 sq. km**.
- State having highest forest cover as percentage of its area is **Mizoram** (88.93 per cent).
- Increase in carbon sink in this report is **103 million tonnes** CO_2 equivalent.
- Top five states with maximum forest cover are

Name of state	Forest cover (in km²)
Madhya Pradesh	77,462
Arunachal Pradesh	67,248
Chhattisgarh	55,586
Maharashtra	50,628
Odisha	50,354

- Top five states / UTs with maximum forest cover as percentage of their own geographical area are as follows:

Name of state / UTs	Forest per cent of geographical area
Mizoram	88.93
Lakshadweep	84.56

A&N islands	81.84
Arunachal Pradesh	80.30
Nagaland	78.21

- Distribution of forest cover in Altitude zones are given below:

Altitude zones	Per cent of forest found
0-500 m	52.50
500-1000 m	28.27
1000-2000 m	10.88
2000-3000 m	5.76
3000-4000 m	2.46
Above 4000 m	0.13

- Out of the total forest cover, the maximum share is of Moderate Dense Forests, followed by Open Forests. The very dense forests in India are in just around **2.5%** of total geographical area of the country.
- Among all the states of India the states which have shown considerable improvement in their forest cover are **Tamil Nadu, Jammu and Kashmir, Uttar Pradesh, Kerala and Karnataka**.
- The states where forest cover has decreased substantially are **Mizoram, Telangana, Uttarakhand, Nagaland and Arunachal Pradesh**.
- The total forest cover in the hill districts of the country is **283,015 sq. km** which is **39.99 %** of total geographic area of these districts. In the latest report the hill districts have recorded a net increase in forest cover of **1680 sq. km area**.
- North east constitutes only **7.98 %** of geographical area of the country but it occupies **one fourth of the forest cover**. However, according to the current report there is a decrease in the forest cover in the north east by **628 sq. km**, which is primarily because of shifting cultivation and increase in biotic pressure.
- In world's total mangrove vegetation, India's share stands at **3 %**. Currently Mangrove cover in India is **4740 km²** which is **0.14 %** of the country's geographical area. Sundarbans in West Bengal accounts for almost half of the total area. As compared to **2013** there is a net increase of **112 sq. km** in the mangrove cover.

- Top five states with maximum Mangrove cover are as follows:

Name of state	Mangrove cover (in km^2)
West Bengal	2106
Gujarat	1107
Andaman & Nicobar Island	617
Andhra Pradesh	367
Odisha	231

- The total carbon stock has also increased by **103 million tonnes** or an increase of **1.48** in percentage terms compared to previous assessments. The total carbon stock in the country's forest is around **7, 044 million** tonnes. The increase in the carbon stock shows the commitment of the country towards achieving INDC target of additional carbon sink of **2.5 to 3.0 billion tonnes** of CO_2.
- The degraded forest lands which have a Canopy density of less than 10 % are called **Scrubs**. The Lands with Canopy density of 10-40% are called **Open Forests**. The Land with forest cover having a canopy density of 40-70 % is called the **Moderately Dense Forest (MDF)**. The Lands with forest cover having a canopy density of 70% and more are called **Very Dense Forests (VDF)**.

ISFR, 2017-

- India among Top Ten Nations in the World in terms of Forest area. The countries are: Russia, Brazil, Canada, US, China, Australia, Congo, Argentina, Indonesia, India.
- India is placed 8th in the list of Top Ten nations reporting the greatest annual net gain in forest area.
- 24.4% of land area under forest and tree cover.
- ISFR is a biennial publication of FSI, an organization under MoEFCC, GoI.
- ISFR 2017 is the 15th in the series since 1987.

The report for the first time contains:

- Information on decadal change in water bodies in forest during 2005-2015
- Forest fire

- Production of timber from outside forest
- State wise carbon stock in different forest types and density classes
  - The total forest and tree cover is 24.39 per cent of the geographical area of the country. In this forest cover is 21.54% only.
  - An increase of 8021 sq.km (about 80.20 million hectare) in the total forest and tree cover of the country, compared to the previous assessment in 2015. In that, 6778 sq.km increase is in forest cover and 1243 sq.km in tree cover.
  - Top three states/UTs having the largest forest cover in terms of area are: Madhya Pradesh (77414 sq.km), Arunachal Pradesh (66964 sq.km) and Chhattisgarh (55547 sq.km)
  - Top three states/UTs in terms of percentage of forest cover with respect to the total geographical area are: Lakshadweep with (90.33 per cent), Mizoram (86.27 per cent) and Andaman & Nicobar Island (81.73 per cent).
  - The total mangrove cover stands at 4921 sq.km and has shown an increase of 181 sq.km. The top gainer in terms of mangrove cover is Maharashtra (82 sq.km).
  - The extent of bamboo-bearing area in the country has been estimated at 15.69 million ha. There has been an increase of 1.73 million ha in bamboo area compared to 2011 assessment.
  - Striving towards achieving NDC goal India is striving towards achieving its NDC goal of creating additional carbon sink of 2.5 to 3.0 billion tonnes of CO2 equivalent through additional forest and tree cover by 2030.
  - As per present assessment total carbon stock in forest is estimated to be 7,082 million tonnes. There is an increase of 38 million tonnes in the carbon stock of country as compared to the last assessment (ISFR, 2015).

CHAPTER 21

Memory based JRF Paper 2015

1. The scientific name of Asiatic lion is- **Panthera leo persica.**
2. During nitrification Nitrogen is converted into nitrite by- **Nitrosomonas.**
3. The maximum mangrove cover is found in- **Sunderbans delta.**
4. Cutting of trees into logs is called- **Bucking.**
5. Science of determining past climate from trees primarily on the basis of annual rings is called- **Dendroclimatology.**
6. The classification of trees according to Champion and Seth is based on- **Ecosystem approach.**
7. Which among these is non-coppicer tree- **Cedrus.**
8. In which form factor, basal area is measured at breast height-**True form factor.**
9. Piece of plant tissue used to initiate a culture is called- **Explant.**
10. Which law is used to estimate the proportionate of trees in selective forests-**De Liocourt'law.**
11. Smallest permanent working plan unit in India is called- **Compartment.**
12. A thinning done in regular crop in anticipation of suppression is called- **German thinning.**
13. The felling in hills is generally to be carried out from- **Uphill side.**
14. Volume of logs in India are calculated by Hoppus rule, which is- **$(g/4)^2 \times l$.**

15. Distribution and representation of age and size classes of trees in a stand is- **Stand table.**
16. Katha is also obtained from a woody climber- ***Uncaria gambier.***
17. Tending operation in forestry does not include- **Regeneration felling.**
18. 3rd World Agroforestry Congress was held in Feb 2014 in the city- **New Delhi.**
19. *Hippophe rhamnoides* is an important plantation species of- **Cold Desert.**
20. Banni grassland are located in- **Kutch area, Gujarat.**
21. Vivipary is found in- ***Rhizophora* spp.**
22. The Run of Kutch is famous for- **Asiatic lion.**
23. A good seed year is observed in *Abies pindrow* once every- **7-8 years.**
24. Allelopathy is best example for- **Amensalism.**
25. The non-random differential reproduction of genotype is called- **Mutation.**
26. Selection which provides rapid genetic gain while simultaneously maintaining the broad genetic base is- **Mass selection**.
27. In paper making process, sizing is done to render- **Impervious to ink**.
28. Transforming the lives and landscapes is the motto of- **ICRAF, Nairobi.**
29. A shelter belt in general protects the leeward area about- **20 times the height of shelterbelt**.
30. Which of the following is not a part of trinity of norms in forest management- **Site quality.**
31. A systematic association between a fungus and the roots of a vascular plant- **Mycorrhiza.**
32. Boucherie process is used for the treatment of- **Green bamboos only.**

33. Soil, leaching or removal from upper horizon to lower horizon is called- **Illuviation.**
34. Indian institute of remote sensing is located at- **Bangalore.**
35. Taungya system originated in- **Myanmar.**
36. Bamboos are- **Grasses.**
37. *Tamarindus indica* is native to- **Africa.**
38. In India, the concept of joint forest planning and management was first started in- **West Bengal.**
39. The total geographical area under forest cover in the world is- **33 %**.
40. Seedling seed orchards have originate from- **Seedlings.**
41. First cultivated crop in the world are- **Wheat & Barley.**
42. Contribution of agriculture to GDP of India is- **14%.**
43. Forest laws are- **Constitutional law.**
44. Teak defoliator is- ***Hyblaea puera***.
45. D&D survey in agroforestry was proposed by- **J.B Raintree.**
46. The type of thinning is done, based on rule of thumb, in early stages of the crop is-**Mechanical thinning.**
47. The biological diversity act was enacted in the year- **2002.**
48. CVP index was given by- **Paterson.**
49. The living fossil tree is- ***Gingko biloba.***
50. MAI of India forest is- **0.5 m^3/ha.**
51. Working plan is generally prepared for a period of- **10 years.**
52. The project tiger was initiated under the leadership of- **Morarji Desai.**
53. Khus oil is extracted from-***Vetiver zizaniodes.***
54. Wood attacking fungi are forms of- **Saprophytic plants.**
55. The transitional zone between two ecosystems is- **Ecotone.**
56. Species suited for green manuring is- ***Gliricidia sepium.***
57. Hedge row intercropping is an example for- **Agri silviculture.**
58. Alkaloid extracted from *Rauvolfia serpentine* is- **Serpentine.**

59. One m^3 of wood is equal to- **35.315 Cft.**
60. In north Indian agroforestry, crop tree combination, that is common in- **Rice & poplar.**
61. Indian institute of forest management is located at- **Bhopal.**
62. Reserved forest are constituted under- **Indian Forest Act, 1927.**
63. Mother plant from which vegetative propagules are taken for propagation is called- **Ramet.**
64. Sal tree belongs to family- **Dipterocarpaceae.**
65. Scientific name of African elephant is- ***Loxodonta africana.***
66. Sandal spike disease is caused by- **Mycoplasma.**
67. REED is- **Reduced Emission from Deforestation and Degradation.**
68. Diffusion treatment is applicable to- **Green timber with water soluble wood.**
69. Champion and seth classified the Indian forests into- **16 groups.**
70. Aggregate of measurement or totality of individual is called as- **Population.**
71. Most of the commercially important oleoresin are obtained from- **Conifers.**
72. Father of green revolution is- **Norman Borlaug.**
73. In India forestry research was initiated after establishment of FRI in the year- **1906.**
74. A tree which has grown out of its geographical area is - **Exotic**.
75. The maximum genetic gain is achieved through- **Seedling Seed Orchard.**
76. Crown thinning is also called as- **French thinning.**
77. Which among the following is an annual flowering bamboo- ***Arundinaria wightiana.***
78. A measure of the relative productivity of a site for a particular species is- **Site quality.**

79. Removal of only certain sp. of high value of trees above a certain size and of certain sp. Without full regard to silvicultural requirements is called- **Selective felling.**

80. Number of plants required for 10 ha of plantation in which the plants are planted in a triangular pattern 5 m will be- **4620.**

81. The interval between seedling felling and final felling is called- **Rotation period.**

82. Separation of fibres as seen both on the cross selection and longitudinal surfaces of the material is known as- **Shake.**

83. Curvature both across width and along length is called- **Cupping.**

84. The fibre is extensively used for making elephant harness- ***Sterculia villosa.***

85. A valuable red dye is obtained from the roots and stems of- ***Rubia cordifolia.***

86. World agroforestry centre is located at- **Nairobi, Kenya.**

87. The amendment which brought forests and wildlife in the concurrent list in VII schedule of the constitution was- **42th amendment**.

88. The progressive yield was advocated by- **Hartig.**

89. A rotation in which all the items of revenue and expenditure are calculated with compound interest is- **Financial rotation.**

90. Schneider's formula for increment percent is- **P= 400/nD**.

91. The genotypical response of a species ton a particular habitat is called- **Ecotype.**

92. If in narrow sense heritability is O, there will be- **No environmental variation.**

93. A situation in which the vegetative propagule does not assume tree form but continues to grow like a branch is- **Plagiotrophic growth.**

94. The causal organism of casuarina stem wilt is- ***Trichosporium vesiculosum.***

95. The cost that docs not vary with the level of production is- **Fixed cost.**

96. Oligopoly means- **Few sellers.**

97. An oldest agroforestry practice in humid tropical climate of India is- **Home garden.**

98. Shade loving plants are called- **Sciophytes.**

99. Water movement from high conc. to low concentration is known as- **Diffusion.**

100. The presence of VAM enhances the uptake of nutrients of- **Phosphorus.**

101. Agrostology deals with – **Grasses.**

102. Juglan is an allelochemicals of- **Walnut.**

CHAPTER 22

Memory based JRF Paper 2016

1. Hybridoma technology is used in the production of
   (a) Hybrid plants (b) **Monoclonal antibodies**
   (c) Monoclonal serum (d) Polyclonal antibodies
2. Spike disease in sandal is caused by
   (a) Fungi (b) bacteria
   (c) **Mycoplasma** (d) Insect
3. Forest conservation Act was formed in
   (a) 1927 (b) 1827
   (c) 1985 (d) **1980**
4. Latest forest policy was formulated in the year
   (a) 1898 (b) 1894
   (c) **1984** (d) 2011
5. World environment day is celebrated on
   (a) 5th July (b) 11th July
   (c) 16th September (d) **5th June**
6. Biodiesal plant is
   (a) ***Jatropa*** (b) *Madhuca*
   (c) *Azadirachta* (d) *Aloe*
7. Offence and defence challan in the forest act comes under the section
   (a) **Chapter ix** (b) Chapter xi
   (c) Chapter x (d) Chapter v
8. Transition zone between two communities called as
   (a) Climax zone (b) **Ecotone**
   (c) Edge (d) Succession

9. Monoclimax theory was given by
   (a) E. P. Odum (b) A. G. Tansley
   (c) Reiter (d) **Clements**
10. In forestry D.B.H is measured at the height
    (a) 1.37 cm (b) 1.30 cm
    (c) **1.37 m** (d) 1.30 m
11. In the slope area D.B.H is measured from
    (a) Low hill side (b) **uphill side**
    (c) Both sides (d) None
12. Which N_2 base is found only in RNA
    (a) Adenine (b) Guanine
    (c) Thymine (d) **Uracil**
13. Which cell organelle have stroma and grana lamellae
    (a) Nucleus (b) **Chloroplast**
    (c) Ribosomes (d) Endoplasmic reticulum
14. Ozone depleting factor is
    (a) Carbon di oxide (b) Sulphuric acid
    (c) Nitric acid (d) **Chloro floro carbon**
15. Alternate source of fuel energy is
    (a) CNG (b) Biogas
    (c) Propane (d) **All of these**
16. Indian Lac Research Institute is situated in
    (a) Jhansi (U.P) (b) **Ranchi**
    (c) Bihar (d) Maharashtra
17. ICFRE was formed in the year
    (a) **1901** (b) 1919
    (c) 1929 (d) 1927
18. Indian Institute of Forest management is situated in
    (a) Jaipur (b) Dehra Dun
    (c) Ranchi (d) **Bhopal**
19. Indian Institute of Plantation management is situated in
    (a) Kerala (b) **Bangalore**
    (c) Tamil Nadu (d) Haryana

20. The shape of a typical shelter belt is
   (a) **Triangular** (b) Square
   (c) Conical (d) Cylindrical
21. Table based on one variable-
   (a) **Local volume table** (b) Regional volume table
   (c) General volume table (d) Yield table
22. One example of height measuring instrument-
   (a) Barometer (b) **Spiegel Relaskop**
   (c) Hygrometer (d) Dendrometer
23. The canopy range of dense forest is
   (a) **Less than one** (b) More than one
   (c) Equal to one (d) None
24. Geo-stationary satellite revolved at a distance of
   (a) **800 m** (b) 1200 m
   (c) 6400 m (d) 22000 m
25. *Foris* is a
   (a) Greek word (b) **Latin word**
   (c) English word (d) Sanskrit word
26. What are the factors that causes uprooting of seedlings
   (a) Wind (b) Rainfall
   (c) Temperature (d) **Both a and b**
27. Particle density is higher than
   (a) **Bulk density** (b) fluid density
   (c) Real density (d) Soil density
28. Soil having pH 3.2 is
   (a) **Acidic** (b) Alkaline
   (c) Basic (d) Acidic-alkaline
29. Transfer of pollen from anther to stigma of one flower to another flower of same plant
   (a) Autogamy (b) Allogamy
   (c) Geitanogamy (d) **Both a and c.**
30. Noise pollution is measured in,
   (a) Hertz (b) **Decible**
   (c) Amplitude (d) Centimetre

31. The RNA which carry information is called
    (a) m-RNA (b) **t-RNA**
    (c) r-RNA (d) hn-RNA
32. In tree damping off is caused by
    (a) **Fungi** (b) Virus
    (c) Bacteria (d) Earthworms
33. The name of teak defoliator is
    (a) ***Hyblea pupurea*** (b) *Hyblea shorea*
    (c) *Hyblea tectonae* (d) *Hyblea foligae*
34. Removal of vegetation from an area and development of newly vegetation is called
    (a) Afforestation (b) Plantation
    (c) **Reforestation** (d) Deforestation
35. Selection of parents on the basis of performance of progeny is called
    (a) Mass selection (b) **Progeny test**
    (c) Family selection (d) Sib selection
36. Plant and animal is differentiated by having
    (a) Ribosome (b) Mitochondria
    (c) **Cell wall** (d) Nucleus
37. Formation of a variety after a sudden change in genetic material of a cell is called
    (a) Hybrid (b) **Mutant**
    (c) Sibling (d) Half sibling
38. Cross of F_1 with either of two parents is called
    (a) Test cross (b) Pure cross
    (c) **Back cross** (d) None
39. The type of root in *Heritiera* is
    (a) Tap root (b) **Pneumatic roots**
    (c) Adventitious root (d) All of these
40. The formula for Von Mental is
    (a) **2NGS/r** (b) 4NGS/r
    (c) NGS/2r (d) NGS/4r

41. The cow unit for camel is
(a) 1 (b) 2
(c) 4 (d) **8**

42. Home garden is suitable for
(a) **Humid tropical** (b) Tropical region
(c) Arid region (d) Semi-arid region

43. The cutting of tree upto browsing is called
(a) **Pollarding** (b) Trainning
(c) Prunning (d) Canopy management

44. Example of bad coppicer
(a) ***Cedrus*** (b) *Shorea*
(c) *Eucalyptus* (d) *Toona*

45. Alley cropping is also called
(a) Home garden (b) **Hedge row**
(c) Jhum cultivation (d) Taungya system

46. The angle between two adjacent teeth is called
(a) Kerf (b) Face
(c) **Pitch** (d) Gullet

47. NRCAF is located at
(a) **Nairobi** (b) Rome
(c) Geneva (d) Switzerland

48. Morgis is the headquarter of
(a) ICFRE (b) **IUCN**
(c) ICAR (d) WWF

49. Jhum cultivation is a common practices in
(a) **North-east** (b) Western
(c) Southern (d) Eastern-west

50. Age of tree can be calculated by
(a) **Annual rings** (b) Tree branching
(c) Stand table (d) Volume table

51. Compact ring is shown by
(a) **Fast growing** (b) Slow growing
(c) Moderate growing (d) None

52. The increment in volume of fast growing species is
(a) **60 cm^3** (b) 120 cm^3
(c) 10 cm^3 (d) 100 cm^3

53. Prokaryote have
(a) Nucleus (b) Mitochondria
(c) **Genophore** (d) None

54. Example of non-leguminous N_2 fixation tree is
(a) **Casuarina** (b) Pongamia
(c) Teak (d) Sal

55. The gestation period of rhinoceros is
(a) 240 (b) 280
(c) 400 (d) **450**

56. *Hipophae rhamnoides* is found in the region of
(a) **Cold desert** (b) Tropical Humid
(c) Arid region (d) Hot desert

57. Bakarwal tribes is residing in which state of India
(a) **Jammu & Kashmir** (b) Rajasthan
(c) Uttarakhand (d) Assam

58. Local control is not used in
(a) RBD (b) **CRD**
(c) LSD (d) SPT

59. Scientific name of Red panda is
(a) *Panthera uncia* (b) ***Ailurus fulgens***
(c) *Panthera leo* (d) *Gazella gazella*

60.

a. Soil testing	a. **Troug**
b. Marble	b. **Limestone**
c. Clay	c. **<0.002 mm**
d. Law of minimum	d. **Leibig**
e. Hue	e. **Dominant spectral colour**

61.

a. Gurgaon Project	a. **F.L. Bryne**
b. Sevagram Project	b. **M.K. Gandhi**
c. Sriniketan Project	c. **R.N. Tagore**
d. Marthandam Project	d. **Hatch**
e. Etawah Project	e. **Albert Mayer**

62. a. Kaziranga National Park — a. **Rhinoceros**
b. Sunderbans tiger reserve — b. **Tiger**
c. Dachigam National Park — c. **Hangul**
d. Gir National Park — d. **Asiatic Lion**
e. Dudhwa National Park — e. **Tiger**

63. a. Green revolution — a. **Rice and wheat production**
b. White revolution — b. **Milk production**
c. Blue revolution — c. **Fish production**
d. Yellow revolution — d. **Oilseeds production**
e. Red revolution — e. **Tomato/meat production**

64. a. World Biodiversity Day — a. **May-22**
b. World Environment Day — b. **Jun-05**
c. World Oceans Day — c. **Jun-08**
d. World Population Day — d. **Jul-11**
e. International Tiger Day — e. **Jul-29**

65. a. Very light timber — a. **300 kg/m^3**
b. Light timber — b. **301-450 kg/m^3**
c. Moderately heavy timber — c. **451-600 kg/m^3**
d. Heavy timber — d. **601-800 kg/m^3**
e. Very heavy timber — e. **801-950 kg/m^3**

66. a. *Mallotus philippensis* — a. **Kamela dye**
b. *Bixa orellana* — b. **Annatto dye**
c. *Butea monosperma* — c. **Dhak dye**
d. *Indigofera species* — d. **Blue dye**
e. *Lawsonia inermis* — e. **Heena dye**

67. a. Succession in fresh water — a. **Hydrosere**
b. Succession in salty water — b. **Halosere**
c. Succession in acidic water — c. **Oxalosere**
d. Succession in dry region — d. **Xerosere**
e. Succession on rock — e. **Lithosere**

68.	a. Damping-off	a. ***Fusarium***
	b. Root disease	b. ***Ganoderma* sp.**
	c. Pink disease	c. ***Corticium salmonicolor***
	d. Spike disease	d. **Mycoplasm**
	e. Heart-rot	e. **Fomes fungi**
69.	a. Poisson distribution	a. $0 \leq x \leq \infty$
	b. Correlation coefficient	b. **-1 to +1**
	c. f-test	c. $0 \leq x \leq \infty$
	d. Regression coefficient	d. $-\infty$ **to** $+\infty$
	e. Probability	e. **0 to 1**

CHAPTER 23

Memory based JRF Paper- December, 2018

1. Hybrid
1. Hormone which is responsible for photoperiodism - Florigen
2. Cultch dye obtained from which tree species- Acacia catechu
3. Brazilian dye obtained from which tree species- *Caesalpinia sappan*
4. Tapping done in which tree spp- Pine
5. Saline soil pH- <8.5
6. Aggregate fruit example- Pineapple
7. Ecosystem term given by- A.G Tansley
8. Type of Bamboo fruit- Caryopsis
9. Which tree species used for best quality paper making- Eucalypts
10. Tree species which is suitable for coastal areas plantation-Casuarina
11. TCIP stands for- Tree cultivation Incentive Programme
12. Why Khejri tree (Prosopis cineraria) regeneration decrease- Pod eating
13. R.B.D is used for- Field experiment
14. C.R.D is used for- Lab experiment
15. United carpels called- Syncarpous
16. World environment day celebrated- 5th June
17. Which nutrient leached in soil- K
18. 3rd Worlf Agroforestry seminar was held- New Delhi
19. Green house gas- CH_4
20. ASCU is which type of wood preservative-
21. Scientific wood preservation started at?

22. Perfect correlation value range- -1 or +1
23. DBH measure in UK at- 1.30 m
24. Similar triangle height measurement instrument- Improved callipers
25. What is pollarding- Pruning at which animal can't browse
26. In clear sky light which type of frost produce- Radiation frost
27. Mycorrhiza provides which elements to plants- P
28. Flowering time of Neem- May-June
29. Wind breaks how much % wind speed decrease- 25 to 75%
30. What is cull- Unmerchantable portion of log
31. Concentration of CO2 in Atmosphere- 350 ppm
32. 1 hectare equal to- 10,000 m2
33. What is alley cropping- Agriculture crop gown in the between of line of tree
34. What is farm forestry- Grown tree n farm boundary along with crops
35. What is Race- Genetically different individual of same species
36. Period of Neem Seed viability- 3 months
37. L.C.C Suitable for Agroforestry System- I to IV
38. Spruce and Fir tree are- Coniferous species
39. Circumference of log- 2 pie r
40. Seedling felling done in which silviculture system- Clear felling system
41. Example of TBOs (Tree born oil seeds)- Pongamia pinnata
42. What is main objective of tree improvement- Production of superior tree
43. Application of outcross in forestry- To increase disease resistance
44. Maximum carbon stock found in- Grassland ecosystem
45. Definition of carbon sequestration- Carbon elimination from atm to plant body
46. Based socio economic Agroforestry classified into- 3 categories
47. Agroforestry originates from which system-Taungya system
48. Maximum MAI in Popular in which age- 8 years
49. In Lower gigantic plain eroded soil cover which tree suitable for this- Acacia species-

50. Volume table based on one variable- Local volume table
51. Formulae for calculation of volume- Von Mental formula
52. What is “Gompa” in Himachal- Traditional Agroforestry system
53. Which nutrient losses when old trees removing-Sulphur
54. Taungya is introduce in India by- Brandis
55. Why do shifting cultivation- For food production
56. What should be natural regeneration condition- Adequate amount of seed in soil
57. What is the benefit of creeping fire in pine forest- Help in natural regeneration
58. Most common method of veneer making- Chip making
59. Veneer logs makes from- Plywood
60. Cutting of small trees called as- Thinning
61. Close nutrient cycle occurs in which ecosystem- Tropical rain forest
62. What is seed orchard- A place of quality seed production
63. Why decrease wildlife population- Due to habitat loss
64. Home garden not found in which area- Arid zone
65. How to rise Shisham quality plantation- Stump planting
66. Main competition in Agroforestry- Tree crop
67. What is NABARD- Agriculture funding bank
68. Establishment year of NABARD- 29th July, 1982
69. Family of Bamboo- Poaceae
70. IIFM located at- Bhopal, M.P.
71. Apiculture with tree known as- Other AF systems
72. NBPGR located at- New Delhi
73. Home garden suitable in which region/area- Humid region
74. What eat queen honey bee- Royal Jelly